Nifty E-Z C
D-STAR Op

Another guide in the
Nifty! Ham Accessories
Easy Guide Series

www.niftyaccessories.com

Copyright

Disclaimer and Limitation of Liability

Contents

About This Guide **1**

Special Thanks To **2**

Chapter 1: D-STAR **3**
History 3
D-STAR Overview 4
D-STAR's Bits and Bytes 8
Repeater System Configuration 9
Programming D-STAR Call Sign Parameters 12
Using D-STAR Gateways 13
Operating Simplex 14
Local / Same Repeater Operation 15
Local Cross-band Repeater Operation 16
Repeater Node Routing 17
Call Sign Routing 19
Setting the UrCall field back to CQCQCQ 21
One-touch Reply 22
Automatic Call Sign Update Prevention 23
Multicast Groups 24
Identify Where You Are Calling From and Wait 25
Limiting Position Beaconing and Data Mode Operation 26

Chapter 2: Dplus Gateway Operation **27**
Dplus Gateway Linking 28
Establishing a Dplus Gateway Link 29
Dplus Reflector Linking 30
Establishing a Reflector Link 34
Local Simulcast 35
Echo Audio Quality Testing 36
Checking Repeater Link / ID Status 37

Chapter 3: Gateway User Registration **39**
Getting Registered 39

Chapter 4: Setting Up Call Sign Memories 41
Call Sign Memories 41
Editing the Call Sign Routing Register 42
Copying from UrCall, Repeater and MyCall Memory Banks 43
Programming UrCall, Repeater and MyCall Memory Banks 45
Programming Your Own Call Sign 47
Recalling Call Sign Fields from a Frequency Memory 48
Organizing D-STAR Repeater Calling Modes in Memory 48
Received Call History 51
Examining Calls in the Received Call Memory 51
Copying Calls from the Received Call Memory 54

Chapter 5: DV Short Text Messaging 57
Programming DV Short Messages 57
Reviewing Short Messages 60

Chapter 6: Internet Resources 61
D-STAR Routing and Linking Calculator 61
Operating the D-STAR Calculator Program 61
jFindu Repeater Locator and Last Heard Lists 65

Chapter 7: Radio Programming Software 67
Icom's Programming Software 67
RT System's Programming Software 68
Icom's RS-91 and RS-92 Programming Software 68
D-STAR Operation Using the RS-92 Software 71

Chapter 8: DV Mode Slow-speed Data 75
D-STAR Oriented Data Communication Software 76
Radio / PC Configuration for Low-speed data Operation 77
Configuring Serial Ports 77
Automatic / PTT Data Transmission Selection 78
Disabling GPS Mode Transmission 79
d*Chat Application Installation and Setup 80
d*Chat Program Operation 83
D-RATS Application Installation and Setup 85
D-RATS Program Operation 87
Configuring and Sending D-RATS QST Messages 89
Transferring Files with D-RATS 90

File Transfer Problems....................91
Using and Creating D-RATS Forms....................92
Stations and Sessions Tabs....................93
Other D-RATS Capabilities....................93

Chapter 9: DV Dongle, D-STAR Adapter....................95
Computer System Requirements....................96
Installing the DV Dongle Software on Your Computer....................96
Selecting the DV Tool COM port and Audio Devices....................97
Setting the Headset and Microphone Audio Levels....................99
DV Dongle Operation....................101
Connecting to Repeaters Linked to a Reflector....................101
Dongle LED Status Indicators....................102
Installation Problems....................102

Appendix: D-STAR Web Pages....................103

About This Guide

Using easy to understand language and illustrations, this guide describes how the D-STAR system operates and provides guidance for setting up your transceiver to be able to access D-STAR's many features and modes of operation. We will go light on theory, concentrating instead on the practical issues of getting things programmed and making voice and digital data contacts.

D-STAR is an evolving technology. Thanks to improvements made by Icom and the effort of many hams creating and maintaining programs such as Dplus, d*Chat, and D-RATS, D-STAR's communication capabilities are far improved from several years ago. The creation of the DV Dongle, which enables worldwide communication without using a radio, has added a whole new dimension to D-STAR operations.

In early 2009, when this book was written, the software running on most gateways was Icom's G2 program supplemented by Dplus version 2.2. No doubt, future enhancements will continue to provide more exciting new communication capabilities.

Lets get started!

Special Thanks To

We wish to thank all those that helped in the creation of this book. Special thanks to Icom who materially supported the project with technical help and generously allowed us to use the graphics from various Icom publications. Ray Novak, N9JA, Icom's Amateur Radio Division Manager was especially helpful in providing contacts that were of assistance in completing the project. Fred Varian, WD5ERD, with Icom Technical Support not only answered my many questions, but also reviewed a draft copy of this book.

We are also indebted to Cecil Casillas, WD6FZA, administrator and champion of the Southern California PAPA repeater system who supported the project by answering my questions and allowing me access to their excellent system of DSTAR repeaters. Without their support I would have been unable to perform the testing and experimentation necessary to verify many of the DSTAR features and procedures presented in this book.

Several other PAPA system members were also supportive of my efforts. Allen Klisky, KB6OYA answered questions and helped me run tests using digital mode operation with the d*Chat and D-RATS programs. Ted Petrina, W6SAT and Craig Davis, KM6AW both took of their valuable time to review draft copies of the book, providing me with corrections and suggestions.

Chapter 1: **D-STAR**

Hams have a long history of applying digital technology to amateur radio communications. Starting with RTTY, a succession of other digital modes has ensued: Packet Radio, PSK, PACTOR and many others. D-STAR is the latest and perhaps most comprehensive effort to date, offering reliable digital voice and data communication all over the world.

History

After three years of research, the D-STAR protocol was published by the JARL (Japanese Amateur Relay League) in 2001. The research to investigate digital technologies for use in amateur radio was funded by the Japanese government and undertaken by a committee of Japanese radio manufacturers and interested observers. Icom, the primary promoter of this new technology, provided the equipment used for the development and testing phase of the program.

At first, adoption of the technology outside of Japan was relatively slow. However, in the last several years D-STAR repeater systems have started coming into their own. With the increasing availability of D-STAR repeater systems and gateways, the numbers of hams using these systems is showing dramatic growth.

D-STAR repeaters and gateways are now available in many areas of the United States, Europe, Canada, South America and Australia. Repeaters linked to Internet Gateways provide voice and data communications all over the world.

To encourage equipment suppliers to adopt the technology, JARL published the D-STAR protocol as an "open" specification that details the over-the-air interface and repeater/gateway transport requirements for interoperability of D-STAR equipment. To date, Icom is the only manufacturer of D-STAR capable repeater systems and radios. As the technology becomes more widely adopted, other manufacturers may chose to offer equipment as well.

D-STAR Overview

D-STAR (Digital Smart Technologies for Amateur Radio) offers digital voice and slow and high-speed data communications. The slow-speed digital voice and data is transported at 4800 bps, with 3600 bps being used for voice and voice error correction, the remaining 1200 bps is used for synchronization and general use. Of this 1200 bps, about 900 bps is available for transporting data. High-speed digital data communication is transported at 128 kbps, supports Ethernet packets, and is fast enough for interactive Internet applications.

By connecting repeater sites over the Internet forming, a world-wide radio network, the D-STAR system provides state-of-the-art functionality to amateur radio repeater systems.

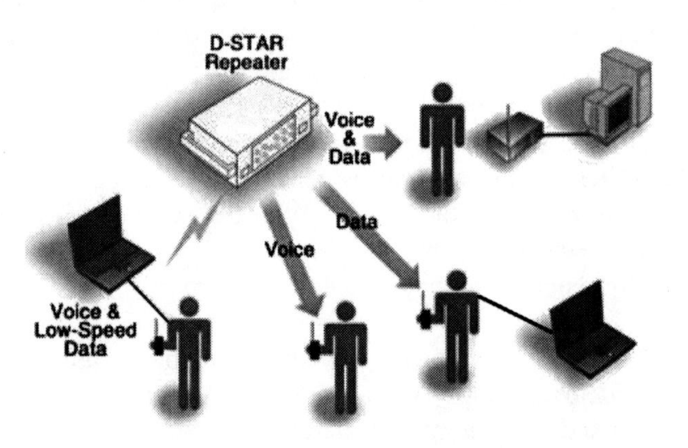

In D-STAR, voice communication is referred to as DV mode (digital voice) operation. Voice is converted to a digital format using an electronic chip called a CODEC, which encodes and decodes audio signals in the AMBE (Advanced Multi-Band Excitation) format.

To the critical ear the audio quality of a D-STAR voice signal might sound slightly inferior to a high quality FM signal, but is more than adequate for intelligible voice communications.

The nice thing about digital voice operation is that the quality of the signal remains crystal clear until it is lost. As long as the signal remains above a minimum threshold, it can be decoded without degradation and will remain clear without the path noise or "picket fencing" weak signal artifacts common on traditional FM mode communications. If the signal falls below the level required for decoding, communication will drop out or become garbled, sounding a bit like the R2D2 Star Wars character.

At first, operating D-STAR is a bit unnerving. After years of using conventional FM repeaters, its strange not to hear a squelch tail after releasing PTT. D-STAR repeaters drop the carrier almost immediately upon releasing PTT on the transceiver; consequently the momentary squelch tail hiss that we are accustomed to is not there. Being conditioned to delay transmission until after you hear a courtesy beep and then operating on a repeater without a beep can throw you off. Even though D-STAR repeaters don't broadcast courtesy beeps, it's still important to pause before replying, as it gives other stations a chance to break in. Not to worry though, after using D-STAR a bit that strange feeling soon goes away, being replace by the thrill of using this new mode of communication.

Interestingly, in DV mode slow-speed 1200 bps digital data can be transmitted at the same time, and on the same frequency while you are engaged in a voice conversation. Since both voice and data are being handled digitally, they can be transmitted together on the same signal without any interference to your voice conversation.

Don't be misled by the term slow-speed, 1200 bps DV mode data is more than capable of keeping up with typing on a keyboard and for transmitting short messages and small amounts of data. Subtracting out header and message blocking overhead, DV mode data has about 900 bps available for general use and is much faster than PSK31, but slower than 9600 bps packet operation. Like packet, DV mode data is unsuitable for sending large files or "surfing the web."

In addition to the slow-speed DV data that can be transmitted simultaneously with your voice on the 144, 440 MHz and 1.2 GHz bands, D-STAR supports a high-speed digital data rate of 128k bps on the 1.2 GHz band. Due to packet overhead and other factors, actual throughput is closer to 90k bps. Referred to as DD mode (digital data), this high-speed data capability is unique in amateur radio because it is fast enough to support exchanging large files, pictures and for user-interactive Internet e-mail and web browser applications.

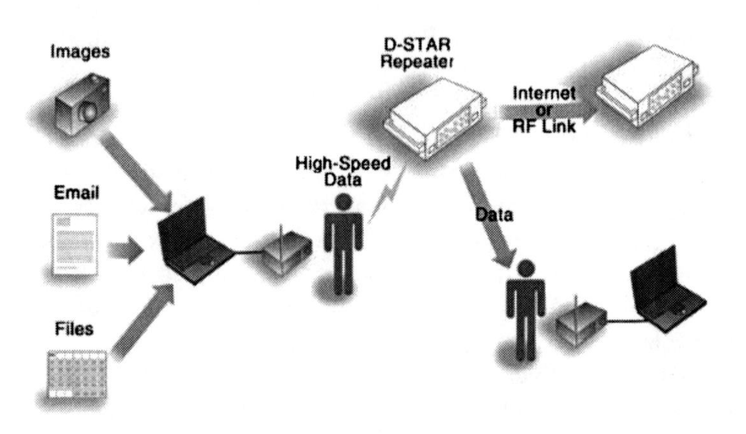

Connecting your PC, laptop or PDA is simply a matter of connecting a cable to the radio, no external TNCs or other devices are required. For slow-speed data, depending upon the radio, either an RS-232 serial or USB cable is used. High-speed data connections are made using a standard Ethernet cable. Low-speed data capabilities are built into all VHF / UHF D-STAR transceivers currently being supplied by Icom. High–speed data is limited to radios with 1.2 GHz capabilities.

For emergency communications, one of the advantages of D-STAR's digital data capabilities is that messages can be locally transported independent of the Internet when the "lines are down".

But wait, there is more! A single repeater or a group of repeaters can be connected to the Internet via a device called a gateway and are referred to as a "Zone." Gateways use the Internet to connect to other D-STAR gateways and reflectors anywhere in the world. This allows you to communicate to hams located in areas far removed from your local repeater, somewhat similar to IRLP operation but with an interesting added capability. Whenever you key-up, your call sign is automatically transmitted via the digital transport mechanism built into the radio. When the gateway routes your call, it also stores your call sign locally and provides it to the Internet connected *Trust*

Server. In this way the D-STAR system keeps track of which repeater you were last heard on.

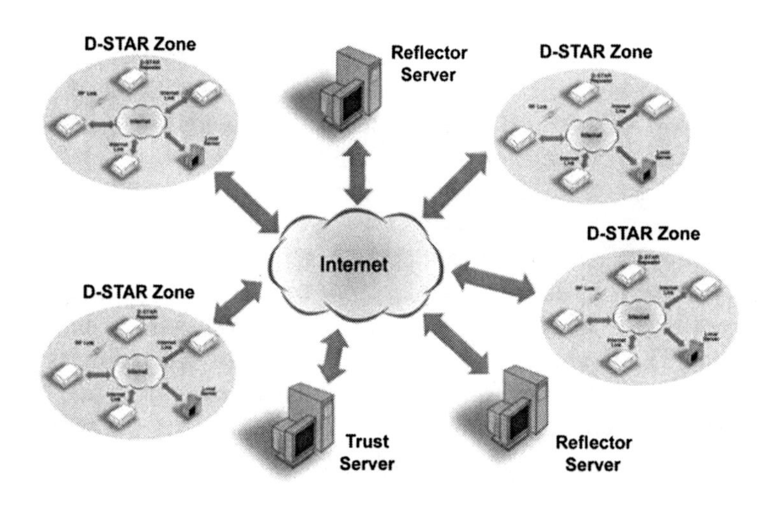

By entering the call sign of whom you want to contact into your radio, you can make a directed call to that specific ham. The technique is referred to as *Call Sign Routing* and unlike IRLP, you don't need to know which repeater he is on. Periodically all gateways synchronize their local data with data located on the *Trust Server*. The gateway system uses that data to figure out which repeater your friend was last heard on and automatically routes your call to that repeater. *Call Sign Routing* can be thought of as being similar to how a cell phone operates. As you travel around, the cell system "knows" where you are at and directs incoming calls to the cell tower nearest to your location. D-STAR works much the same way.

With *Call Sign Routing,* after entering the call sign of the person you are trying to reach, the D-STAR system can automatically route your call to other repeaters even if they are on a different band or in a different city. As a result, no matter which repeater your friend might have switched to, your call will be routed to where he was last heard. This solves the problem of having to make calls on all the repeaters that your friend might frequent.

D-STAR's Bits and Bytes

D-STAR DV mode (slow-speed digital and voice) transceivers produce an RF signal that is quite different than those produced by conventional FM transceivers. The voice portion of the output signal is not FM modulated; audio is directly converted to a digital data stream using a AMBE (Audio Multi Band Encoder) codec chip, in turn the AMBE voice data is combined with other digital data to form a simultaneous composite voice and digital data stream, which is then transmitted as a GMSK modulated signal.

Within the D-STAR specification, the exact format of this composite digital stream is defined as the *Common Air Interface*, or *CAI* protocol and is made up of a *Radio Header* followed by the *data* payload. The *Radio Header* consists of a series of synchronizing and control bits followed by four call signs used to route the signal to its intended destination. The data payload portion consists of alternating *Frames* of *Voice* and *Data* information: a frame of 72 bits of voice followed by a frame of 24 bits of data, a pattern which continuously repeats until followed by a unique termination frame of 48 bits. This pattern of alternating digital voice and data frames occurs regardless if there is voice and no data, or if there is data and no voice. Space in the payload is always reserved for the voice and data frames regardless of whether they are used or not.

For those interested in the detailed structure of the *Common Air Interface* protocol and other technical details of the D-STAR over-the-air protocol, an English copy of the JARL specification can be downloaded from
www.jarl.com/d-star/shogen.pdf

A more comprehensive look at the D-STAR over-the-air protocol is provided by Peter Loveall, AE5PL in his excellent paper titled *D-STAR Uncovered.* This paper provides additional insight and information beyond what is in the JARL specification, including a summary of Icom's enhancements to the base specification.
http://www.aprs-is.net/downloads/DStar/DSTARUncovered.pdf

Both of the above documents are quite technical, describing the air-link communication protocol in exacting detail and are the basis for how the system is designed. These documents are primarily of interest to those designing D-STAR compatible equipment or software.

If this bits and bytes stuff is all Greek to you, don't worry, a detailed understanding of the underlying voice and data transmission protocol is not essential for enjoying the benefits of D-STAR operation. It's really no different than driving modern automobiles, using computers or accessing the Internet, all of which rely on complex systems and do not necessitate our detailed understanding before being able to successfully operate them.

Repeater System Configuration

As shown below, a D-STAR repeater system consists of one or more repeater modules that are interconnected via a repeater controller, which also supports a connection to the Internet via a PC running a D-STAR compatible gateway program under the LINUX operating system.

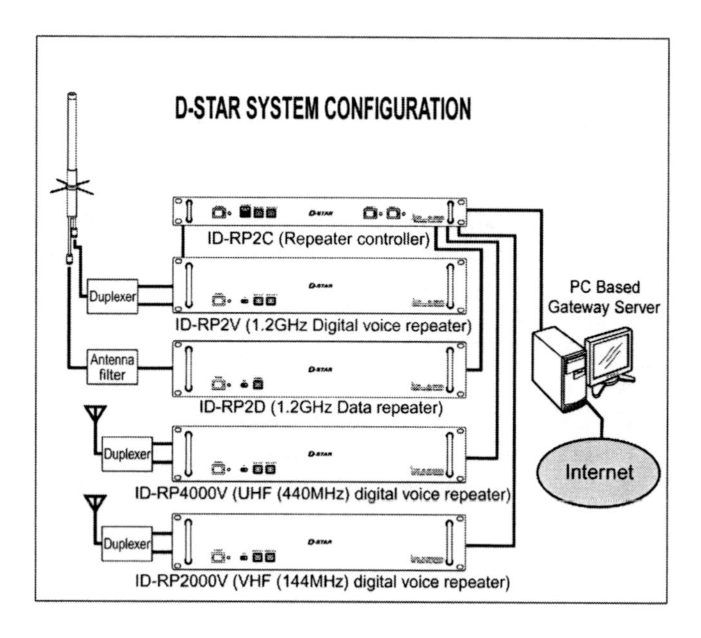

A repeater system can be configured with repeater modules supporting digital voice on the 144 MHz, 440 MHz and 1.2 GHz bands. A given installation may include any combination of the three voice modules. Notice that if high-speed digital data is to be supported, a separate 1.2 GHz digital data repeater module is required.

A repeater system can be configured without including a PC gateway server to the Internet. Of course, the repeater system loses the capability of communicating with remote gateways and repeaters, but still provides functionality similar to that provided by conventional stand-alone FM mode repeaters.

It's common practice to refer to individual repeater modules connected to a repeater controller in a D-STAR system as *nodes*, *modules* or *ports.* For example, the four repeater modules shown in the preceding diagram can alternately be referred to as *nodes*, *modules* or *ports*. Regardless of which term is used, they all refer to a specific repeater module. The term port is derived from the practice of referring to individual repeater modules as being connected to a repeater controller's ports.

The agreed upon practice for naming these modules is to add a letter designating the individual port after the main call sign for the system. Regardless of the length of the main call sign, the port designation is always placed in the 8th character position, preceded by as many spaces as necessary to fall in the 8th position.

The "_" underline characters below are only used to illustrate the required spaces. When entering an actual call, use "real" spaces not the underline.

W6XYZ__A	DV	1.2 GHz voice repeater, Port A
W6XYZ__A	DD	1.2 GHz high-speed data repeater, Port A
W6XYZ__B	DV	440 MHz band data / voice repeater, Port B
W6XYZ__C	DV	144 MHz band data / voice repeater, Port C
W6XYZ__G		The system's Internet gateway, Port G

To avoid confusion as to which local repeater and *port* you are operating through, always indicate the *port* you are on. For example, when making a call through W6XYZ, you would announce “N6FN (using your own call sign) calling on W6XYZ port B.”

Notice that the entire repeater system is given a single call sign. Repeater call signs are restricted to a maximum of 6 characters. Individual repeater modules are identified via the port designation suffix that appears in the 8th character position. Unlike conventional repeaters, D-STAR repeater systems must have unique call signs, they cannot be an individual’s call sign, otherwise *Call Sign Routing* would not be able to function.

Programming D-STAR Call Sign Parameters

Programming a transceiver to make calls (or for linking to a gateway or reflector) involves programming call signs into the four parameters of the *Call Sign Routing Register*:

- **UR CALL** Call of the station, node or reflector you are calling.
- **RPT1** Call of the local repeater node you are calling from.
- **RPT2** Call of a destination repeater or your local gateway.
- **MY CALL** Your own call sign, or call sign variations.

Different transceiver models may display slightly different abbreviations for these four parameters, but on all radios they accomplish the same thing. These parameters are programmed in different ways depending upon how you are making the call: simplex, local repeater, Repeater Node or Call Sign Routing, and Gateway or Reflector linking.

UR CALL This is either the station you want to talk to, or is set to **CQCQCQ** so you can call CQ or work round-table as on a conventional FM repeater. When using a gateway, **UR CALL** is used to designate the call sign of the individual you are calling, or it can also be used to control gateway linking or for accessing a reflector.

RPT1 Used to enter the call sign of the local repeater you are using. The 8th character position is special as it specifies the band and port you are operating on. You need to insert spaces as necessary to make sure that the port switch letter (A, B or C) falls in the 8th character position.

RPT2 This is the call sign of where we want our transmission to go, either to one of the other ports on the same repeater system, or to the gateway used to access the D-STAR network. Again, the 8th character position is special because it is used for the port switch designation letter: A, B, C or G.

MY CALL This is used for your own call sign, or perhaps a variation of it with a suffix indicating different radios that you might be using.

In this book, the *Call Sign Routing Register* refers to the call sign memory that the radio uses to make D-STAR calls. As we will see in Chapter 4, depending upon the radio, there are several ways for making entries into the *Call Sign Routing Register.*

- Manually editing the current *Call Sign Routing Register*
- Copying from the **UrCall**, **Repeater** and **MyCall** memory banks
- Using the [**RX-CS**] key one-touch reply feature
- Recalling a memory channel that has these parameters set

Generally you would not want to be programming call signs whenever you wish to call someone. And, you certainly don't want to attempt programming call signs while driving. The normal practice is to store in advance all call and repeater node combinations you expect to use into call sign memories within the transceiver. Once your transceiver has been programmed, it's a simple matter of recalling the correct call sequence from memory.

Local DSTAR repeater groups typically provide the information required for accessing your local repeaters. In addition, as we shall see in Chapter 6, there are several web-based resources providing call sign information for D-STAR systems all over the world.

Using D-STAR Gateways

One of the key features of D-STAR is the ability to communicate with other D-STAR systems over the Internet. Indeed, connecting repeater systems via gateways is one of the most powerful aspects of D-STAR operation. Most of the enhanced capabilities of D-STAR repeater systems rely on their gateway connection.

Since so much can be done via the gateway system, the creators of D-STAR have implemented a worldwide gateway user registration system in order to prevent misuse of the resource. Users must register to be able to operate any D-STAR features that involve gateway access to the Internet. Without registration you are generally limited to simplex and local repeater operation.

Except for the "Operating Simplex" and "Local / Same Repeater Operation" sections below, you will need to register for gateway access if you wish to try any of the features described in the following sections

Operating Simplex

D-STAR transceivers are capable of working station-to-station on simplex, just like conventional FM transceivers. One advantage of a D-STAR radio is that it's already equipped for digital communications. This can simplify situations when you want to transmit data, perhaps for emergency communications or public service events such as aid stations spread along a bike or foot race. If high-speed communications are required, two 1.2 GHz, ID-1 transceivers can transfer data directly without the use of a repeater.

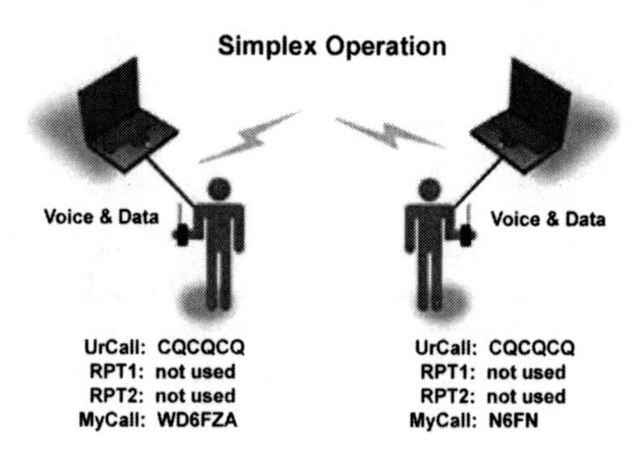

To talk to anyone on simplex without having to input their call sign, the **UrCall** field is programmed with **CQCQCQ**. Since repeaters are not used when working simplex, depending upon the radio, **RPT1** and **RPT2** are programmed as "not used" or left blank. And of course, your own call sign is used in the **MyCall** field.

Here we see that N6FN and WD6FZA, except for their individual call signs, have their radios set the same way and will be able to talk to anyone on their simplex frequency.

Local / Same Repeater Operation

When working locally on a single D-STAR repeater module, the call sign of the local repeater module is used in **RPT1**, and depending upon the radio, **RPT2** can be marked as "not used" or left blank.

Single / Local Repeater Operation

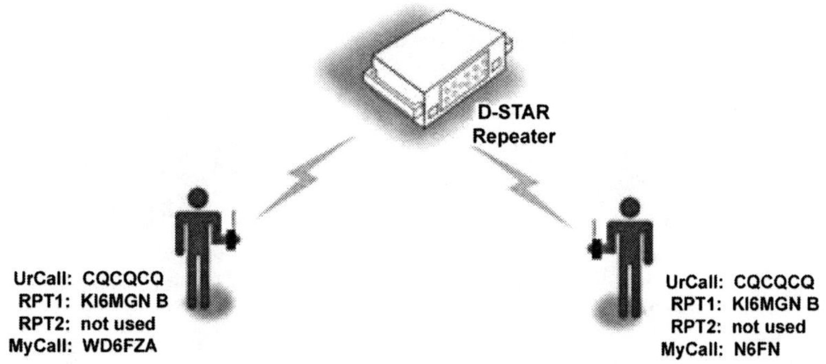

D-STAR
Repeater

UrCall: CQCQCQ
RPT1: KI6MGN B
RPT2: not used
MyCall: WD6FZA

UrCall: CQCQCQ
RPT1: KI6MGN B
RPT2: not used
MyCall: N6FN

In the above example notice that the call sign for **RPT1** is KI6MGN_B, which indicates it is a 440 MHz repeater attached to port B of the controller. Using **CQCQCQ** in the **UrCall** field allows inter-communication between all users on the repeater without having to enter a specific station's call sign.

With the call signs set as shown, operation is very similar to a conventional FM repeater with everyone being able to hear each other and participate in the conversation.

Note: So that linked gateways, reflectors and DV Dongle users can hear your traffic, most D-STAR system administrators recommend that the **RPT2** field be set to your local gateway. In the case of the PAPA system KI6MGN repeater, **RPT2** would be set to KI6MGN_G.

Local Cross-band Repeater Operation

If your local repeater system has two or more modules, you can work cross-band just as if you were operating on a single repeater. In this case, **RPT1** specifies which repeater you are operating through, and **RPT2** specifies the destination repeater. The local repeater controller takes care of routing the signals between the two ports.

Local Cross-Band Operation

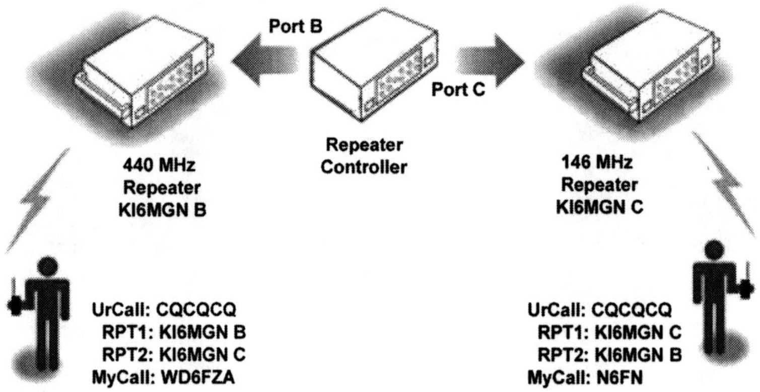

Port B

Port C

Repeater Controller

440 MHz Repeater KI6MGN B

146 MHz Repeater KI6MGN C

UrCall: CQCQCQ
RPT1: KI6MGN B
RPT2: KI6MGN C
MyCall: WD6FZA

UrCall: CQCQCQ
RPT1: KI6MGN C
RPT2: KI6MGN B
MyCall: N6FN

In this example WD6FZA is going through the KI6MGN_B, 440 MHz repeater, and N6FN is going through the KI6MGN_C 146 MHz repeater. It is important that the port switch designation (the letters B and C in this case) is programmed into the 8th character position.

Notice that since the two stations involved are on separate repeater modules serviced by the same repeater controller that a gateway is not being used and that the call signs they programmed into **RPT1** and **RPT2** of their radios are reversed.

When the station hearing the call, in this case N6FN, wants to respond he needs to set his radio's **RPT2** field to the radio module being used by the calling station, in this case KI6MGN_B. But keep in mind that the calling station needs to identify which module he is on so the answering station can configure his radio to the repeater module that the calling station is using.

Repeater Node Routing

Repeater Node Routing, also called "Source Routing," "Port Linking" or a "Zone Call" allows the user to specify a specific repeater node as the destination for his transmission. This can be used to place a call to a specific ham or perhaps as a way of announcing your presence or calling CQ on a distant repeater.

Using this method a user can either send his signal to a different port on the same repeater system or to any gateway connected repeater node in the world.

Repeater Node Routing

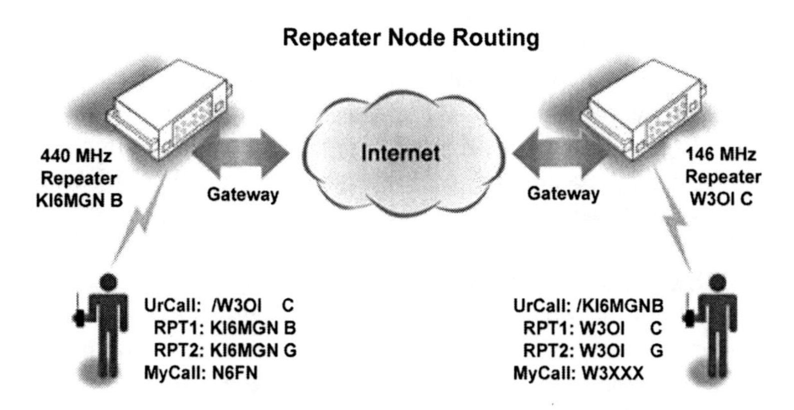

To use Repeater Node Routing a " / " is placed in front of the destination repeater's call sign in the **UrCall** field. The leading " / " character lets the controller know you are making a call to a specific repeater node and that it's not the call sign of a person.

Note: After N6FN made his call to the W3OI__C repeater, W3XXX configured his radio using Node Routing to route his call back to N6FN's repeater, KI6MGN_B. The "G" suffix in the 8th character position of the **RPT2** field indicates that the signal is to be routed to the gateway.

When someone answers a call made using Node Routing, they must configure their radio to route their signal back to the repeater module that the source radio is using. Therefore, as is generally the case when using D-STAR, the calling station needs to identify which repeater and port he is calling from.

A receiving station, in addition to hearing the transmitting station identify the repeater he is calling from, can also examine the *Received Call* memory on his radio. Refer to the *Received Call History* procedure in Chapter 4 for details on how to examine the *Received Call* memory.

When you are finished with your QSO on the remote repeater you need to change the **UrCall** field back to **CQCQCQ**, otherwise when making any further contacts, even on your local repeater, your voice will still be routed to and heard on the remote repeater indicated in the **UrCall** field. This is an easy mistake to make and I suppose everyone has done it at one time or another.

I supposedly know better, but here is how easy it is to make a mistake. Hearing Toshi JF1CXH, a Japanese station, calling, on our local D-STAR repeater, I configured my radio for Node Routing back to his repeater by placing his local repeater call sign, /JP1YIQA into my radio's **UrCall** field. So far so good and the QSO went fine.

The problem "snuck in" after my QSO with Toshi in Japan ended. Just as I signed off with Toshi, I was immediately called by another station on my local repeater. Not thinking, I returned his call and we chatted a bit about making contacts to Japan and a few other topics. Only after finishing the follow-on QSO did I remember that I had neglected to switch my **UrCall** field back to **CQCQCQ**. The result was that my side of the follow-on conversation was broadcast in Japan! I was the cause of several minutes of unintended interference on their repeater. Not good!

This can happen when using any of the D-STAR modes where you are either routed to or connected to a remote repeater or reflector. Its important to remember to reconfigure your radio's **UrCall** field and if necessary also the **RPT1** and **RPT2** fields back to where they need to be to prevent "interference" on a remote repeater node. Stay alert and don't let this happen to you.

Call Sign Routing

With *Call Sign Routing*, also referred to as *User Linking,* you can make a call to a specific ham without having to know what repeater system he is on. Providing a ham has registered for gateway operation, whenever he transmits on a gateway-equipped system, the D-STAR gateway system's database is updated with the repeater system module he was last heard on.

When you make a directed call to a specific amateur's call sign, the gateway system automatically routes your call to the repeater module where that station was last heard.

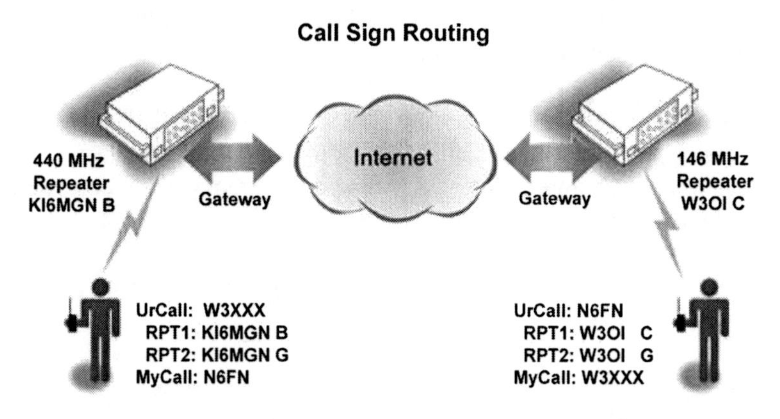

In this example *Call Sign Routing* is being used by N6FN to locate and talk to W3XXX, which happens to be on the W3OI___C module at the moment. Notice that W3XXX has set his **UrCall** field to N6FN's call sign to answer his call. The other three fields were probably already set since he has recently been using the W3OI___C repeater.

Before the ham you are calling can respond, he needs to program his radio's **UrCall** field with your call sign. The responding station can manually enter your call into his radio (or select it if already has it in memory) or he can use the [**RX-CS**] "one-touch" key on his radio, if it has one. The more recent D-STAR capable transceivers have the one-touch capability to copy a received station's call sign to the **UrCall** field.

Using the one-touch reply [**RX-CS**] key only temporarily copies N6FN's call sign into the **UrCall** field, and will last until something else is placed into the **UrCall** field. It's not permanently saved anywhere. One-touch operation is described in the following section.

Note that both stations have set **RPT2** to their local gateway. Doing so has allowed the gateway system to rout N6FN's call to the last repeater module that W3XXX has been heard on. As a side note, it also allows Dongle users to hear both sides of the conversation.

One of the issues with *Call Sign Routing* is that the "last heard on" process can take an hour or more to update the database; therefore the user may no longer be on that repeater.

One way to partially get around the problem, so that you can immediately receive calls when away from your local repeater, is as soon as you are on another D-STAR repeater system place a call back to your local repeater. Then at least your local repeater system will immediately know what repeater you are on, and if anyone calls you from there using *Call Sign Routing*, their call will be forwarded to where you are at.

Of course, if your friend is using **CQCQCQ** in the **UrCall** field when operating on your local home repeater, you will not receive any calls directed to you unless he knows you are out of town and switches over to *Call Sign Routing* by entering your call sign into his **UrCall** field.

By the way, if out of town, what method do you use to call back to your home repeater system? You have a couple of choices: you could either use *Call Sign Routing* if you wanted to call a particular station, or if you just wanted to check in with a general call on your home repeater you could use *Repeater Node Routing*.

When you are done talking with your friend you need to change your **UrCall** field back to **CQCQCQ**, otherwise when making any further contacts, even on your local repeater, your voice will still be routed to and heard on the remote repeater where your friend was last operating.

Setting the UrCall field back to CQCQCQ

As we have seen, its important to change your radio's **UrCall** call sign routing field back to **CQCQCQ** after setting it to something else for making calls to specific stations, repeaters or sending linking commands as we will see in the next chapter.

Setting UrCall field to CQCQCQ on the IC-91AD and IC-92AD:

1. Starting with the DV operating mode selected and a D-STAR repeater frequency being displayed on the screen.
2. Press [**0/CQ**] until you hear a beep, then release.

Setting UrCall field to CQCQCQ on the IC-2820:

1. Starting with the DV operating mode selected.
2. If necessary, press [**F**] twice to access the DV mode function keys. (**CS CD CQ R>CS** etc.)
3. Press [**CQ**] to set the **UR** field to **CQCQCQ**.

To set **CQCQCQ** on other radios, refer to the Icom user manual for your radio.

One-touch Reply

The one-touch feature, available on newer model Icom transceivers that have the [**RX-CS**] key, is a handy way of responding to a call. As calls are received they are automatically stored in the *Call History* memory and are available for use by the one-touch feature. Pressing the [**RX-CS**] key sets the radio to respond to the most recent called received.

However, if another called is received after the one you want to respond to, you will need to select the desired call from the *Received Call History* memory as shown in step 3 below.

Be aware, that if the repeater produces a transmission after the desired call was received, one-touch may copy the repeater's gateway call sign to the **UrCall** field. If so, skip step 2 and use step 3 to select the desired call.

Using One-touch on the IC-91AD and IC92AD:

1. First make sure that your own call sign has been set into the **MY** call field and that **RPT1** and **RPT2** are set for *Call Sign Routing* from your local repeater.
2. (Only use Step 2 or 3) After a call has been received, press and hold the [**CALL/ RX-CS**] key one second to set the *Call Sign Routing Register* to respond to the most recently received call.
3. **Or** if you want to select a call sign from a list of recently received call signs, press and hold the [**CALL/ RX-CS**] key and rotate [**DIAL**] to select the desired call sign record. Recently received call signs stored in the *Call History Memory* are displayed at the bottom of the screen.
4. Everything is now set; press [**PTT**] to transmit.

Using One-touch on the IC-2820:

1. First make sure that your own call sign has been set into the **MY** call field and that **RPT1** and **RPT2** are set for *Call Sign Routing* from your local repeater.
2. After a call has been received, press [**F**] twice to access the DV mode function keys. (**CS CD CQ R>CS** etc.)
3. (Only use Step 3 or 4) Press [**R>CS**] key to set the *Call Sign Routing Register* to respond to the most recently received call.
4. **Or** if you want to select a call sign from a list of recently received call signs, press the [**CD**] key and rotate [**MAIN•BAND**] to select the desired **CALLER** call sign from the **RX CALL SIGN** history memory.
5. Everything is now set; press [**PTT**] to transmit.

Important: After using the one-touch feature to make a reply, be sure to change your radio's **UrCall** field back to **CQCQCQ**.

Automatic Call Sign Update Prevention

Icom's D-STAR radios have two menus that can enable received call signs to automatically replace call signs in the *Call Sign Routing Register*. In general you don't want this to happen, so you should verify that they are turned OFF, which is the default setting for these two menus.

The "RX Call Sign Auto Write" menu should be set to OFF to prevent having received station's call sign automatically replace your **UrCall** setting. The default setting is OFF.

The "Repeater Call Sign Auto Write" menu should be set to OFF to prevent having received station's RPT1 and RPT2 call signs automatically replace your RPT1 and RPT2 settings. The default setting is OFF.

Multicast Groups

Multicast is a feature that Icom added with the G2 version of the gateway software. Multicast allows an administrator to associate a group of repeater nodes with an alias (a name of his choosing). Using a Multicast group name allows an administrator to route transmissions between as many as 11 repeater nodes. Multicast group names of up to seven characters long always start with the character " / ". When this name is referred to, it has the same effect as referencing all of the repeater nodes in the network at once.

UR CALL:	/CA1200	Multicast group name.
RPT1:	WR6BRN_C	Local module you are linking from.
RPT2:	WR6BRN_G	Gateway for local repeater module.
MY CALL:	N6XXX	Your own call sign.

While users don't have the capability of creating Multicast Groups, they can make use of the feature by placing the group name in their **UrCall** field. When a user transmits using a Multicast group name, he will be heard on all the repeater nodes in the group. The user's local gateway accomplishes this by sending a stream of data to each of the nodes in the group.

So that users on remote repeaters can respond, it is important to announce the repeater you are calling from and the network name being used. If responding stations don't use the network name in their **UrCall** field, they will not be heard on all the repeaters in the network.

Important: At the end of the round table or net and going back to normal operation, it is important that each station terminate multicast operation by resetting their **UrCall** field back to **CQCQCQ** or something else. Otherwise, their transmissions will still go out to the entire group of repeaters.

Identify Where You Are Calling From and Wait

Whenever making calls on a D-STAR system it is imperative that you identify the repeater and port you are calling from. Otherwise when a station hears your call he won't know if you are local or elsewhere. If you are not on his local repeater he may need to know where you are transmitting from in case he needs to change the settings in his **UrCall** and **RPT2** call sign fields.

Typically you would state the location of the repeater and which port you are on. For instance if operating on the PAPA system Mount Palomar repeater you might say that you are "using Mount Palomar port B."

This brings up the second point to remember when making calls over the D-STAR system. After making a call, monitor long enough for a responding station to make any radio changes necessary. Since it is likely that the responding station was otherwise preoccupied, it may take a few minutes for him to "put down" what he was doing, change his radio settings and return your call. Repeating your call once or twice (don't get carried away here like calling CQ on HF) may allow the station to make note of where you are at so he can make the required settings.

On repeaters with frequent traffic, you may also want to hold transmission a sufficient amount of time to allow a receiving station to use his "one-touch reply" key to copy your call sign information to his radio's D-STAR call sign fields. Remember, he has to hear your call, pick up his radio and then press the key. If your call is too short, by the time he is ready to press the key your signal may have been pushed down in the stack of calls in the *Call History Memory*. Remember, one-touch retrieves the most recent call from the *Call History Memory*.

Limiting Position Beaconing and Data Mode Operation

Whenever multiple D-STAR repeater nodes are linked together for group, emergency or net operations, via Multicast or the Dplus linking methods described in the next chapter, automatic APRS / DPRS GPS position beaconing should be turned off or be set to transmit only on PTT. Automatic beaconing every few minutes will result in data being transmitted to all connected nodes, causing collisions (doubling) with other user's transmissions.

Likewise, for the same reason, consider delaying DV data mode transmission until the net is over or switch to another repeater system. While data can be sent along with a voice transmission, if automatic data transmission has been selected, it won't wait for a voice transmission to occur. It will occur anytime data is ready to be sent.

Chapter 2: **Dplus Gateway Operation**

Dplus is an auxiliary gateway operating system program used to route calls and establish links between D-STAR repeater nodes. Dplus adds features to gateway operation beyond those provided by Icom's G2 gateway software. Because of its enhanced capabilities, Dplus, written by Robin Cutshaw, AA4RC is almost universally installed on all gateways outside of Japan. If installed on your local gateway, among other things, it provides Gateway and Reflector linking capabilities as well as an Echo Test and a means of verifying link status.

As a side note, the Dplus gateway software is periodically upgraded as new features are added. While the installation and maintenance of the Dplus software is performed by your system administrators, new features, which may affect operation, could be introduced at any time. If so, your system administrator will notify you of any operational changes that may need to be made.

Dplus Gateway Linking

With Dplus installed, it's possible for a user or a system administrator to initiate a link between two repeater modules residing on different gateway systems.

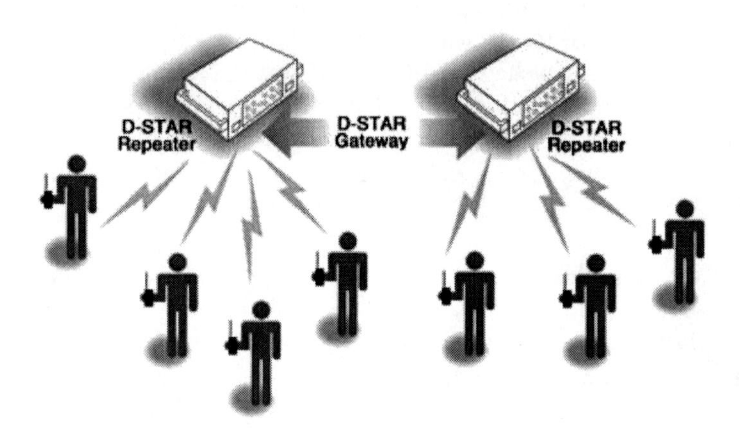

Once a link has been established, all users on the two linked repeater modules can contact each other in a manner similar to conventional FM repeaters that have been linked together. When any station on one repeater transmits, all stations on the remote gateway linked repeater hear that transmission. The system takes care of routing signals back and forth and users should not use any special routing. Their **UrCall** sign configurations should be set to **CQCQCQ** and **RPT2** should be set to the call sign of their local gateway with a "G" in the 8th character position, just as if they were making local calls.

Depending upon the system in your area, you may find repeater nodes linked together temporarily or semi-permanently either by individual users or by a system administrator.

Some systems may not allow individual users to perform linking operations. Establishing a link should not be attempted without first contacting the administrators of a D-STAR system to find out if it is allowed and to determine any procedures they want used.

Establishing a Dplus Gateway Link

Administrators can establish links via their command line control system for the gateway. If enabled on their local repeater system, users can establish a gateway link to another repeater by setting the **UrCall** field to the call of the remote repeater followed by the letter "L" in the 8th character position to signal that you wish to establish a link.

Step 1. Programming the Call Signs for making the link:
For example, if you are currently located on the 2-meter module of the WR6BRN system and wanted to link to the 70cm module of the KI6MGN_B system, the *Call Sign Routing Register* would be set as follows:

UR CALL:	KI6MGNBL	Repeater module B you are linking to.
RPT1:	WR6BRN_C	Local module you are linking from.
RPT2:	WR6BRN_G	Gateway for local repeater module.
MY CALL:	N6XXX	Your own call sign.

Step 2. Key up and identify

Once you have programmed the call fields for making the link, key up and state your intention: "N6XXX, activating link to KI6MGN module B".

Step 3. Change UrCall back to "CQCQCQ"

When you hear a voice announcement saying, "Remote System is Linked." the link is established. Before trying to make any contacts, change your radio's **UrCall** field back to **CQCQCQ**. If you don't do this, the system will attempt to re-establish the link every time you transmit. While this is not catastrophic, it does burden the system with unnecessary activity.

UR CALL:	**CQCQCQ**	Set to communicate with any station.

You might also hear another message: "System Currently Linked," which means that the repeater you are trying to link to is already linked to somewhere else, or you that didn't change your radio's **UrCall** field back to **CQCQCQ**.

Step 4. Un-Link at the end of the QSO

You must remember to unlink the repeaters when you are finished. As a practical matter any station can terminate the link, it does not have to be the originator of the link. To terminate the link, set the **UrCall** field as shown below.

UR CALL: _ _ _ _ _ _ _ U   Place a "U" in 8th character position.

You might also hear another message: "System Not Currently Linked," which occurs if you try and unlink, but the repeater is not currently linked. This can be used to verify that the repeater is indeed unlinked.

Dplus Reflector Linking

Gateways equipped with Dplus can use Internet connected reflectors for linking multiple repeater nodes together, permitting conversations among all users on the linked nodes. The reflector acts as a central conferencing hub, which is the key element that enables linking multiple D-STAR repeater nodes together.

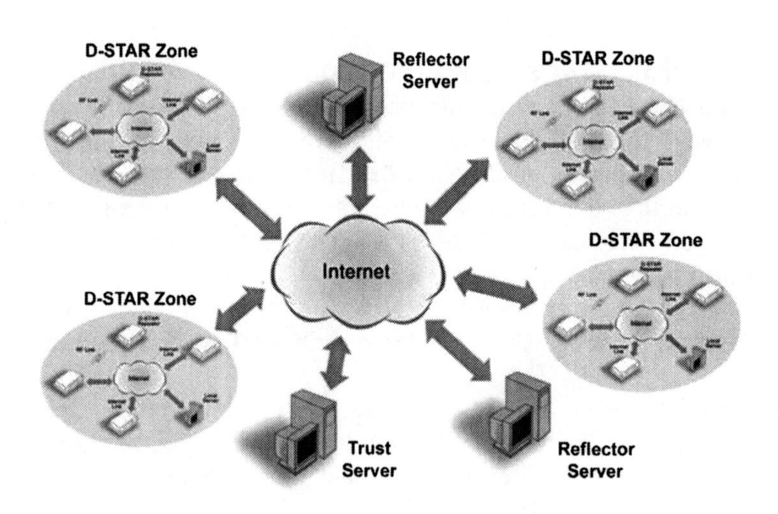

Reflectors are a special type of Internet connected gateway that "reflects" DV mode voice and data back to all linked nodes. When the reflector receives a transmission from one of the linked nodes, it rebroadcasts it to all of the other linked nodes. The effect of this is that all users on all linked nodes are able to hear and talk to each other without having to change their radio's *Call Sign Routing Register* settings. All users should leave their radios set for local repeater operation.

Reflector links are often set up for wide area nets and for emergency communications. Repeaters are generally linked and unlinked from reflectors by system administrators, however some D-STAR systems also allow their users to establish links.

If a Reflector Link has been established, users on any of the linked nodes can talk and hear each other without having to modify their radio's **UrCall** field. The settings normally used for local repeater operation will work, providing the **RPT2** field is set to the call sign of the local gateway with "G" set in the 8th character position as shown below.

UR CALL:	CQCQCQ	Set to CQCQCQ for reflector use.
RPT1:	WR6BRN_C	Local module you are operating on.
RPT2:	WR6BRN_G	Gateway for local repeater module.
MY CALL:	N6XXX	Your own call sign.

Currently there are 17 reflectors worldwide, REF001 through REF017 and more are being added as time goes on. Each reflector has three modules, A, B and C, each of which can support the linking of different groups of repeater nodes together. A reflector's modules can all be active at the same time and are somewhat like having three different conference rooms, each supporting their own group of linked repeater nodes.

A current list of reflectors, where they are located and a short description of what the A, B, C modules are normally used for can be found at: **http://www.dstarinfo.com/reflectors/reflectors.htm**

A sample screen shot of this web page appears below. Only part of the page is shown here, the full page and current reflector status is available at the above URL.

Reflector	Location	Usage
REF001A	Aurora Illinois, United States	Some nets
REF001B	Aurora Illinois, United States	Some nets
REF001C	Aurora Illinois, United States	Permalink Repeaters - General conversation
REF002A	Nebraska, United States	Southeastern US D-STAR Weather Net
REF002B	Nebraska, United States	Some nets
REF002C	Nebraska, United States	Some nets
REF003A	Australia	Ad-hock & Emergency Use - Australia
REF003B	Australia	Permalink for Repeaters, including all WIA Port B Repeaters - Australia
REF003C	Australia	Australian Nets.
REF004A	United States	Alternate for SE D-STAR Weather Net
REF004B	United States	
REF004C	United States	
REF005A	London, England (100Mbps) Usage Guide	UK Nets, Permalink Repeaters
REF005B	London, England (100Mbps) Usage Guide	French Language - Swiss and French users
REF005C	London, England (100Mbps) Usage Guide	
REF006A	London, England (100Mbps) Usage Guide	Scottish Net, Permalink Repeaters

Clicking on a reflector name will bring up a screen showing a list of repeater nodes and dongles that are currently being linked by that reflector. A sample of which is shown on the following page.

Linked Gateways

Module A	Module B	Module C
	KI6MGN C	
	KF6BQK B	
	K8LCD C	
	KW6HRO A	
	K6MDD B	
	KI6KQU B	
	K6IFR B	
	KT7APR B	

DV Dongle Users

Callsign
K6BK
WB8REI
KC8YQL
KI6NHY
KM6AW
WA8YGR
WD6FZA
KI6FNZ
AH6KD

Depending upon your local system's usage, you might find that one or more of your local systems repeaters are linked together on a permanent or semi-permanent basis. It is common practice for some repeaters to maintain a full time link to a reflector. Any transmission on any of the repeaters is heard on all of the repeaters. If you are hearing calls from stations that are not in your local area, chances are your repeater is linked with one or more remote systems. If you think this might be the case, you can use the above-mentioned URL to check the link status of your local system.

Establishing a Reflector Link

If your local repeater system allows users to establish links, here is the procedure. The procedure is exactly the same as for Dplus linking of two repeaters, except the name of a reflector module instead of a call sign is used in the **UrCall** field,

Step 1. Programming the command for linking to a reflector:
For example, if you were currently located on the 2-meter module of the WR6BRN system and wanted to link to the REF001C reflector, the *Call Sign Routing Register* would be set as follows:

UR CALL:	REF001CL	Reflector module you are linking to.
RPT1:	WR6BRN_C	Local module you are linking from.
RPT2:	WR6BRN_G	Gateway for local repeater module.
MY CALL:	N6XXX	Your own call sign.

Step 2. Key up and identify
Once you have programmed the call fields for making the link, key up and state your intention: "N6XXX, linking to Reflector 001 module C".

Step 3. Change UrCall back to "CQCQCQ"
When the link is established, you will hear a voice announcement saying, "Remote System is Linked," you then need to change the **UrCall** field back to **CQCQCQ**. If you don't do this, the system will attempt to re-establish the link every time you transmit.

UR CALL: **CQCQCQ** Set to communicate with any station.

Step 4. At the end of the linked session, Un-link
When the linked session is complete set the **UrCall** field as shown below to terminate the link.

UR CALL: _ _ _ _ _ _ _ U Place a "U" in 8th character position.

If the command was accepted you will hear a voice announcing "Remote System is Un-linked."

Step 5. Change UrCall back to **CQCQCQ**

Local Simulcast

Local Simulcast allows you to simultaneously transmit on all the local repeater system modules attached to a given controller. In a full stack system, there would be three voice modules, one for each of the bands: 2m, 70cm and 23 cm.

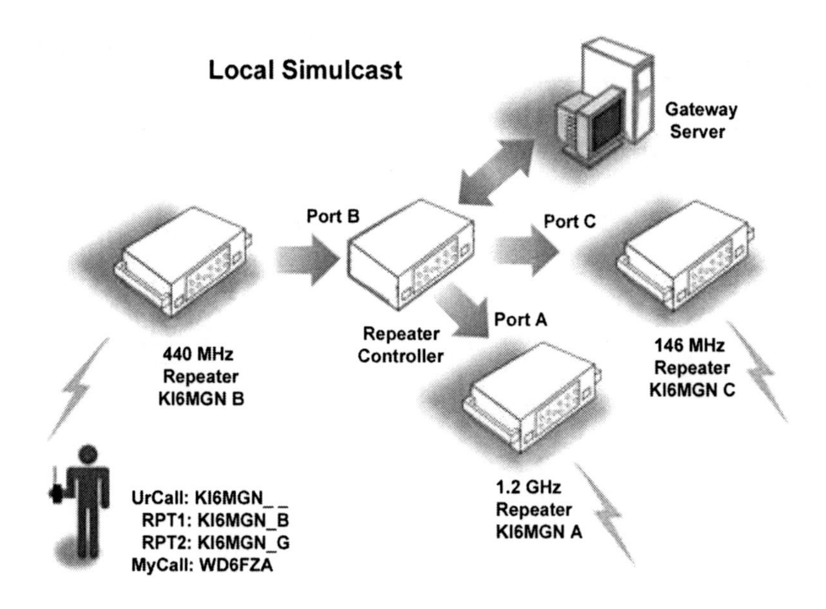

To transmit on all of a repeater system's local modules, program **UrCALL** with the gateway call sign followed by however many spaces are required to fill all 8 positions. Note that in this case, trailing spaces are important, without them the controller will not "know" that this is a simulcast request.

If other stations on the repeater system also have their radios set for simulcast, users on all of the repeater modules can talk to each other.

Echo Audio Quality Testing

Echo audio testing is a handy feature of Dplus equipped gateways. Users can make a short transmission and after unkeying a recording of your transmission will be played back. Besides being a useful check of your audio transmit quality, it can also be used to verify that your local repeater modules and gateway are operating normally. If your system has multiple modules, you could perform a check of each of them using the Echo test. If all is working OK, each module will "hear" your transmission, digitize it and pass it to the gateway server which will delay it a bit and send it right back.

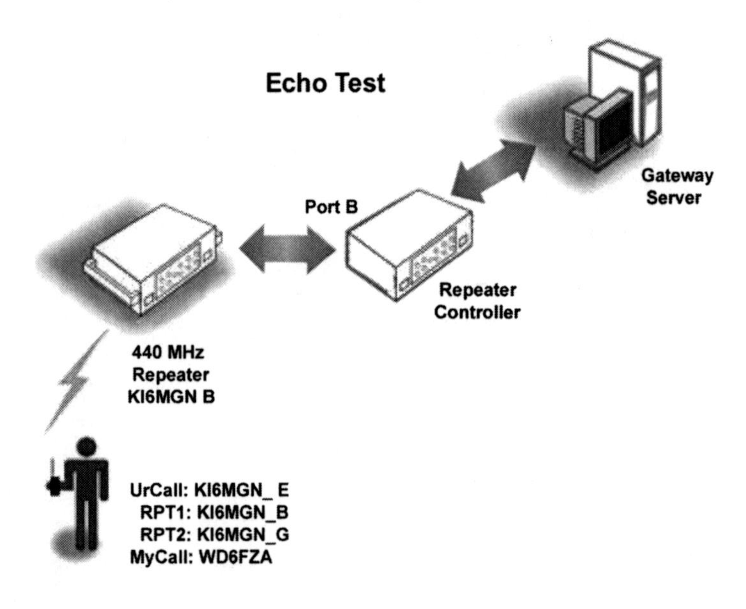

To run an Echo Test, set the **UrCall** field to the call sign of the repeater system you are on, followed by the letter "E" in the 8th character position. The other fields are set as normal for local repeater operation. In this case, **RPT1** is set to port B for testing the 440 MHz module. Other ports on your local system can be checked the same way.

Checking Repeater Link / ID Status

Dplus also comes with the ability to query the repeater to determine if it is linked or not. When the command is transmitted to the repeater, it will respond verbally with the current link status. The exact format of the vocal message may vary on different systems as the system administrator can modify the message sound file. This feature may not be available on all systems.

The link status message may communicate that the repeater module is currently unlinked, or that it is linked to a repeater or reflector.

To perform an ID / Link status check, program the repeater's call sign into the **UrCall** field with the letter "I" in the 8th character position. The remaining call fields are configured as usual for normal repeater operation as shown below.

UR CALL:	WR6BRN_I	ID / Link Status command.
RPT1:	WR6BRN_C	Local module you are operating on.
RPT2:	WR6BRN_G	Gateway for local repeater module.
MY CALL:	N6XXX	Your own call sign.

Chapter 3: **Gateway User Registration**

Registration is required to use the Internet D-STAR Gateway system. Once your registration is approved, the gateways will recognize your call sign and the *Trust Server* will keep track of the repeater nodes you were last heard on. Without registration you are generally limited to local repeater node operation.

Getting Registered

On most systems registration is a two-part process. The first part is providing the required registration information, which will be reviewed by a system administrator. The second part occurs when you receive notification from the administrator that your registration has been approved and you can start using the gateway system. As part of the process, the administrator may also provide you with information for becoming a member of the local repeater organization and instructions for operating the system.

Registration is a simple process, with most local D-STAR systems having provisions for doing it on-line. Contact your local system administrator to find out how they want you to register. Typically all that is required is:

- Your call sign (and perhaps a copy of your FCC license)
- Name
- Home address
- E-Mail address
- If registering on-line, a password for accessing your account

If your local system supports on-line user accounts, after approval you will be given a web page URL for you to log on and setup your account. Initially you will need to enter a call sign for use with your radios. If desired, additional call sign variants may be entered by adding a one-letter suffix, separated by a single space, to your call. Your first entry need not have a suffix appended to your call sign. Call sign variants may be desired for implementing unique *Call Sign Routing* capabilities. *Call Sign Routing* will treat each variant as a unique call sign. Additional variants are required if you will be using more than one I-D1 transceiver in the digital data mode.

For each call sign variant that you enter, the system assigns an IP address used to identify you. An IP address is a unique Internet Protocol identification number assigned to devices communicating on the network. Valid call sign suffix characters consist of all capital letters, (no numbers) except that they may not be placed in the 8th character position, which is reserved for the D-STAR call sign switches: A, B, C D, G, S and I. You need to make an entry for at least one radio to be able to use the system.

Note: Only register once, usually via the registration facilities provided by your local D-STAR gateway system administrators. This will allow you to have access to the entire D-STAR network. Even if you travel to other areas, the repeater gateways will still recognize your call. If you register a second time, either on the same gateway or from any other gateway system, you will confuse the system.

Registration actually occurs on your local gateway. That gateway passes your registration information onto the *Trust Server,* which in turn propagates it to all other gateways in the D-STAR System. Thus when you register on one gateway, you are automatically registered on all the gateways in the network.

If you don't know how to contact the administrator's for your local D-DSTAR systems, you can frequently find out where to register, or at least a website to start from by accessing the Repeater Directory found at **http://www.dstarusers.org/**.

From the D-StarUsers home page click on the *Repeater Directory* link to display a list of all D-Star repeaters. Use this page to find your repeater. Clicking on the repeater's call sign will bring up a variety information about that repeater system, including a web site for the system, and on many of the repeaters a link to the systems on-line gateway registration page.

Chapter 4: **Setting Up Call Sign Memories**

In Chapters 1 and 2 we covered different methods of making calls, and the format of the *Call Sign Routing Register* fields for making those calls. Here we examine how the call sign memories of a typical radio are organized and programmed. We will use the IC-92AD as an example, and also provide programming instructions for the IC-91AD and the IC-2820. Other Icom D-STAR radios have essentially the same functions and can be similarly programmed.

Call Sign Memories

In Chapter 1 we mentioned that the radio uses the *Call Sign Routing Register* to make D-STAR calls. The *Call Sign Routing Register* contains the now familiar set of parameters that are used to control and route D-STAR calls:

- **UR CALL** Call of the station, node or reflector you are calling.
- **RPT1** Call of the local repeater node you are calling from.
- **RPT2** Call of a target / destination repeater or your gateway.
- **MY CALL** Your own call sign, or call sign variations.

Depending upon the radio, there are several ways for making entries into the *Call Sign Routing Register.*

- Manually editing the current *Call Sign Routing Register*
- Copying from *UrCall*, *Repeater* and *MyCall* memory banks
- Recalling a memory channel that has these parameters set
- Using the One-touch feature

Editing the Call Sign Routing Register

On most radios the four call sign memories can be edited "in place." This provides an immediate way of changing one or more of the calls when you don't already have the desired call programmed in one of your call sign memory banks.

Once you have accessed the *Call Sign Routing Register* you can then select individual fields and edit the contents of that particular field.

Call sign editing procedure for the IC-91AD and IC-92AD:

1. Starting with the DV operating mode selected and a D-STAR repeater frequency being displayed on the screen.
2. Press [**MENU**], which brings up the menu list.
3. Rotate [**DIAL**] or use [▲] / [▼] to select **CALL SIGN** entry.
4. Press [▶] to access the **CALL SIGN** *Routing Register.*
5. Rotate [**DIAL**] or use [▲] / [▼] to select the call sign field you wish to edit: **UR**, **R1**, **R2**, or **MY**.
6. Press [▶] to start editing the selected call sign.
7. Use the keys as described at bottom of the screen to edit the selected call sign.
8. When editing is complete, press [**5**/ ↵] to save the call sign.
9. When call sign editing is complete, press [**MENU**] to return to normal operation.

Call sign editing procedure for the IC-2820:

Unlike the IC-91AD and IC-92AD transceivers, on the IC-2820 you are unable to edit the **CALL SIGN** *Routing Register* directly. Instead, to make changes to the **YOUR**, **RPT1**, **RPT2**, or **MY** call sign fields, you must copy call signs from the *Your Call*, *My Call* and *Repeater* call sign memories, which can be programmed or edited as desired before copying. Refer to the following section.

Copying from UrCall, Repeater and MyCall Memory Banks

First lets examine the relationship of the *UrCall*, *Repeater* and *MyCall* memory banks to the *Call Sign Routing Register.*

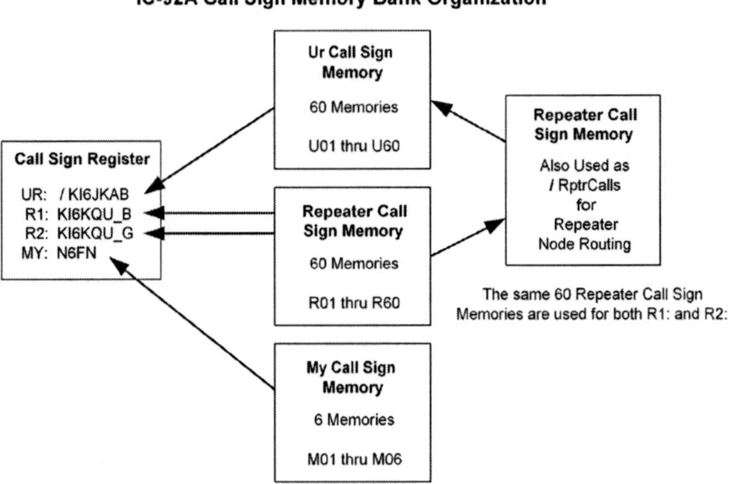

As we see in the diagram, previously saved entries in the *UrCall*, *Repeater* and *MyCall* memory banks can be copied to their respective fields in the *Call Sign Register.* Note that the **R1** and **R2** repeater call fields are both loaded from the same memory bank: the *Repeater Call Sign* memory.

Besides being able to be loaded from the *UrCall* memories, the **UR** field can also use entries from the *Repeater Call Sign* memory. When *Repeater Call Signs* are copied into the *UrCall* field, they are automatically preceded with the " / " character so that they are ready to be used as *Repeater Node Routing* call signs.

Copying Calls to the Call Sign Register on the IC-91AD and IC-92AD:

1. Starting with the DV operating mode selected and a D-STAR repeater frequency being displayed on the screen.
2. Press [**MENU**], which brings up the menu list.
3. Rotate [**DIAL**] or use [▲] / [▼] to select **CALL SIGN** entry.
4. Press [▶] to access the **CALL SIGN** *Routing Register* display.
5. Rotate [**DIAL**] or use [▲] / [▼] to select the call sign field you wish to change: **UR**, **R1**, **R2**, or **MY**.
6. Press [▶] to enter the memory bank for the selected field.
7. Rotate [**DIAL**] or use [▲] / [▼] to select one of the pre-programmed call signs in the memory.
8. Pressing [**5**/ ↵] copies the selected call sign to the register.

Copying Calls to the Call Sign Register on the IC-2820

1. Starting with the DV operating mode selected and a D-STAR repeater frequency being displayed on the screen.
2. Press [**F**] twice to access the DV mode function keys: (**CS CD CQ R>CS** etc.)
3. Press [**CS**] to access the **CALL SIGN** screen.
4. Rotate [**MAIN•BAND**] to select the call sign field you wish to modify: **YOUR**, **RPT1**, **RPT2**, or **MY**.
5. Press [**MAIN•BAND**] and then rotate [**MAIN•BAND**] to select the desired call sign from memory. (Or instead, press [**CQ**] if you wish to enter **CQCQCQ** into the **YOUR** call sign field.)
6. Press [**BACK**] to save your editing and return to the **CALL SIGN** screen, where you can go back to Step 4 and update other call sign fields as required.
7. When done, to exit the **CALL SIGN** screen and return to normal operation, press [**BACK**] as required.

Programming UrCall, Repeater and MyCall Memory Banks

To simplify radio operation, the *UrCall*, *Repeater* and *MyCall* memories should be programmed with all the call signs you are likely to use. Once this is done its a simple matter of copying the desired call sign to the **CALL SIGN** *Routing Register* when you need it.

Programming Call Sign Memories on the IC-91AD and IC-92AD:

1. Starting with the DV operating mode selected and a D-STAR repeater frequency being displayed on the screen.
2. Press [**MENU**], which brings up the menu list.
3. Rotate [**DIAL**] or use [▲] / [▼] to select **CALL SIGN** entry.
4. Press [▶] to access the **CALL SIGN** *Routing Register* display.
5. Rotate [**DIAL**] or use [▲] / [▼] to select the associated call sign field for the memory you wish to edit: **UR**, **R1**, **R2**, or **MY**.
6. Press [▶] to enter the selected field's memory bank.
7. Rotate [**DIAL**] or use [▲] / [▼] to select an empty memory location, or a programmed call sign you wish to overwrite.
8. Press [▶] to start programming the call sign.
9. Use the keys as described at bottom of the screen to enter the desired call sign.
10. When editing is complete, press [**5/** ↵] to save the call sign.

Note: When programming your own call sign, **MY**, you can add a four character note following the " / " symbol. Refer to the following section.

Programming Call Sign Memories on the IC-2820

1. Starting with the DV operating mode selected.
2. Press [**F**] to access the function keys
3. Press [**MENU**] to select the **MENU** screen.
4. Rotate [**MAIN•BAND**] to select **CALL SIGN MEMORY.**
5. Press [**MAIN•BAND**] and then rotate [**MAIN•BAND**] to select the call sign memory you wish to modify:
   a) **YOUR CALL SIGN MEMORY**
   b) **RPT CALL SIGN MEMORY**
   c) **MY CALL SIGN MEMORY**
6. Press [**MAIN•BAND**] and then rotate [**MAIN•BAND**] to select a call sign you want to edit, or an empty memory channel you wish to program.
7. Press [**MAIN•BAND**] to start editing. Use the key functions at the bottom of screen to enter or edit the call sign.
   a) Press [**ABC**] to select between lower and upper case
   b) Press [**12/**] to select between numbers and symbols
   c) Rotate [**MAIN•BAND**] to select characters
   d) Use [**<**] and [**>**] keys to move entry position cursor
   e) Press [**CLR**] to clear selected characters
   f) Press [**CLR**]1 sec to clear all characters after the cursor
   g) Press [**GW**] to turn the gateway setting on and off.
8. Press [**MAIN•BAND**] to save the programmed call sign.
9. To exit the **CALL SIGN MEMORY** screen and return to normal operation, press [**BACK**] as required.

Note: When programming your own call sign, **MY**, you can add a four character note following the " / " symbol. Refer to the following section for suggestions on using the note.

Programming Your Own Call Sign

Use the preceding procedure to program your own call into the *My Call Sign Memory*, which has space for saving up to six call sign variations. If you "registered" any call sign variations when you registered for gateway usage, you may want to program those same variations into the *My Call Sign Memory* bank.

In addition to programming your call sign, you also have the option of programming a four-character "note" which will appear after your call sign. When your call sign is displayed on a receiving station's radio, it is followed by the " / " symbol and the four-character note.

As it is only used for informational purposes, this note is free form; it can be anything that you wish to communicate. The catch is that its only four characters long. A lot of hams use it to indicate the model of radio they are using: IC91, IC92, 2820 and so on. Others, with short names enter their name: Bob, John, Eric, etc. Use your imagination; maybe you can come up with something unusual. Of course, you also have the option of leaving it blank.

Recalling Call Sign Fields from a Frequency Memory

Three of the *Call Sign Routing Register* fields (**UrCall**, **RPT1** and **RPT2**) can be set as a group when recalling a memory channel which has been programmed with a DV mode repeater frequency. For repeaters that you intend to operate on, this is by far the best method to use. Once programmed, operation is simple, you just recall the desired repeater frequency from memory and everything is automatically set.

Programming a memory channel with a DV mode repeater frequency is no different than programming a conventional repeater frequency to memory. Just as with conventional repeater frequency programming, the repeater frequency, mode of operation, in this case DV mode, the power level, and repeater offset are all saved to memory. In DV mode however, the current **UrCall**, **RPT1** and **RPT2** fields of the *Call Sign Routing Register* are also saved to memory. By setting the call sign fields as required for the desired repeater mode of operation, before saving the frequency setup to memory, it is all saved at the same time to a single memory channel.

With a little thought and planning you can get quite clever with this. All the D-STAR modes that you might use on different repeaters can be programmed in advance. Indeed, many local repeater groups can provide you with a "cheat sheet" containing all the standard combinations you might use with their repeaters. Some groups, or your local ham supply store, might be able to either "clone" your radio, or give you a file that can be loaded from a PC with all the standard frequencies and modes already setup. If a file is provided for use with radio programming software, it will be radio specific, so you will need a version of the file that is for your particular radio model.

Organizing D-STAR Repeater Calling Modes in Memory

If you are going to program and organize your own DV mode memory channels, it may be instructive to see how the PAPA system organizes radio memories for accessing their system of D-STAR repeaters. In Southern California we are fortunate to have the PAPA repeater system. With its network of six different D-STAR repeater systems, it provides wide-area coverage over several different counties. With the large number of repeaters in their system, they have worked out a fairly nice system for programming their user's radios.

The PAPA system administrators provide their users with an Icom compatible data file for programming their radios with a standardized set of repeater access configurations. Because of differences in the radios provided by Icom, different .icf formatted files must be used for each model of radio. The IC-91, IC-92, IC-800, IC-2820 etc, all require different .icf files. The PAPA system administrators have a number of different .icf files, called "Code Plugs" to support the different radios.

In addition to providing the programs in the Icom .icf file format that can be downloaded to the radio using Icom's RS programming software, they also provide an Excel spreadsheet compatible .csv file format (comma separated value). Using the .csv format, users can not only print out and examine how repeaters are accessed either locally or via gateways, but with a little work can convert the data to be compatible with other radios or software programs.

If you have access to an .icf file from one radio and want to convert the file to be compatible with a different model radio, a handy conversion program called *D-StarCom* is located at **http://www.d-starcom.com/**. Full instructions and explanation of the file format is also found at the same URL.

At the time this is being written, the standard PAPA system Code Plug contains 79 entries, which go into 79 consecutive frequency memories in the radio. Rather than show all the file entries the PAPA system uses for programming memory channels, the principle of how its organized can be understood by examining just the *Call Sign Routing Register* variations that are used to implement various D-STAR system access modes from one local repeater system.

The PAPA system identifies its six different D-STAR repeater systems using "D" numbers. The numbers range from D1 to D16 with some gaps in the numbering sequence.

The chart below illustrates *Call Sign Routing Register* variations for implementing local repeater operation, echo testing, info request, and several repeater node call configurations to other repeaters within the PAPA system. The comments describe the D-STAR calling / routing methods used when operating from local modules B and C of the D10 repeater. Using this as an example, you can easily create similar configurations for operating on your local system.

Name	UrCall	RPT1	RPT2	Comment
D10B CQ	CQCQCQ	KI6MGN B	KI6MGN G	Local operation on module B, with gateway enabled
D10B ECHO	KI6MGN E	KI6MGN B	KI6MGN G	Audio echo test on module B
D10B INFO	KI6MGN I	KI6MGN B	KI6MGN G	Link status request, module B
D10B> D1	/KI6JKAB	KI6MGN B	KI6MGN G	Repeater node call to KI6JKA module B
D10B> D4	/KF6BQKB	KI6MGN B	KI6MGN G	Repeater node call to KF6BQK module B
D10B>11A	/KW6HROA	KI6MGN B	KI6MGN G	Repeater node call to KW6HRO module A
D10B>11B	/KW6HROB	KI6MGN B	KI6MGN G	Repeater node call to KW6HRO module B
D10B>15B	/KI6KQUB	KI6MGN B	KI6MGN G	Repeater node call to KI6KQU module B
D10B>D10C	/KI6MGN C	KI6MGN B	KI6MGN G	Repeater node call to module C of the local repeater
D10C CQ	CQCQCQ	KI6MGN C	KI6MGN G	Local operation on module C, with gateway enabled
D10C ECHO	KI6MGN E	KI6MGN C	KI6MGN G	Audio echo test on module C
D10C INFO	KI6MGN I	KI6MGN C	KI6MGN G	Link status request, module C
D10C> D1	/KI6JKAB	KI6MGN C	KI6MGN G	Repeater node call to KI6JKA module B
D10C> D4	/KF6BQKB	KI6MGN C	KI6MGN G	Repeater node call to KF6BQK module B
D10C>11A	/KW6HROA	KI6MGN C	KI6MGN G	Repeater node call to KW6HRO module A
D10C>11B	/KW6HROB	KI6MGN C	KI6MGN G	Repeater node call to KW6HRO module B
D10C>15B	/KI6KQUB	KI6MGN C	KI6MGN G	Repeater node call to KI6KQU module B

Note that the Name column in the above chart is the name that is programmed in the memory channel. These are the names that will be displayed on the radio's LCD when performing memory recall operations. The names chosen should help you recall what the individual memory channels are used for. In the case of the PAPA system repeaters, you can see that the "Dnn" designations combined with the **>** symbol indicate the operation to be performed and which repeater modules are involved.

When creating your own set of repeater access memories, by using a similar format you could add node routing to distant repeaters, simulcast and multicast operation, and even repeater and reflector linking operations. Operating different D-STAR modes is greatly simplified when your favorite repeaters and D-STAR modes of operation are saved to memory. Another big plus, once you have verified that they work, they can eliminate errors that otherwise are prone to happen when programming things on the fly.

Received Call History

As calls are received, the radio saves the call signs being used by the station making the call in the *Received Call* memory. This memory has the capacity for storing 20 of the most recently received calls, numbered **RX01** through **RX20**, with **RX01** being the most recent. As new calls are received, older ones are pushed off the bottom (discarded) since they would be numbered beyond **RX20**.

If you need to double-check the call sign of a recent call, the *Received Call* memory can be quite useful. Many times you may only partially hear a person's call sign, or have trouble understanding what the person said. This is especially true when making contacts with foreign countries, where to our ears the person has an accent, which may make it difficult to clearly understand what was said. By checking the *Received Call* memory we can see what the call was.

Examining Calls in the Received Call Memory

The *Received Call* memory stores the person's call sign, the **CALLER**, and also the **RPT1** and **RPT2** call signs that were used.

Examining the Received Call Memory on the IC-91AD and IC-92AD:

1. Starting with the DV operating mode selected and a D-STAR repeater frequency being displayed on the screen.
2. Press [**MENU**], which brings up the menu list.
3. Rotate [**DIAL**] or use [▲] / [▼] to select **RX CALL S** entry.
4. Press [▶] to enter the Received Call Memory bank. You will see the **RX01** memory number and the call sign of the most recently received call, the **CALLER**.
5. Rotate [**DIAL**] or use [▲] / [▼] to select a call record.
6. Press [▶] and then rotate [**DIAL**] or use [▲] / [▼] to view the station's **CALLER**, **RXRPT1** and **RXRPT2** call signs.
7. After examining and making note of the call signs, press [**MENU**] to exit the memory record and return to normal operation.

Examining the Received Call Memory on the IC-2820

1. Starting with the DV operating mode selected.
2. Press [**F**] twice to access the DV mode function keys:
   (**CS CD CQ R>CS** etc.)
3. Press [**CD**] to display the **RX CALL SIGN** screen. You will see the **RX01** memory number and the call sign of the most recently received call, the **CALLER**.
4. Rotate [**MAIN•BAND**] to select a call record: **RX-01** ~ **RX-20**.
5. Press [**MAIN•BAND**] to view the record's **CALLER**, **CALLED** and **RXRPT1** and **RXRPT2** call signs.
6. To exit or view another record press [**BACK**].
7. After examining and making note of the call signs, press [**BACK**] as required to exit and return to normal operation.

When examining *Received Call* records, you may notice that when **RPT1** and **RPT2** are saved they are reversed, placing the gateway call sign in **RPT1**. The following table indicates how calls are saved in the *Received Call* memory.

Calls as used to make the call	Calls as saved in the Received Call Memory	Comment
UrCall: CQCQCQ	CALLER: WD6FZA	
RPT1: KI6KQU_B	RXRPT1: KI6KQU_G	Gateway is in the RPT1 position
RPT2: KI6KQU_G	RXRPT2: KI6KQU_B	
MY: WD6FZA		

Note: Be aware, if you are not going to use one-touch as the method of replying, and instead will be using *Repeater Node* or *Call Sign Routing* program your own local repeater module in the **RPT1** position, and your local gateway call sign in the **RPT2** position (not the caller's local module or gateway from the Received Call memory).

For *Call Sign Routing* place the caller's call sign in the **UrCall** field, which does not require any call sign editing, use the call sign as it was received. The one-touch function can do this for you, if necessary refer to Chapter 1 for one-touch operation.

Repeater Node Routing requires editing the call sign to insert the " / " character in front of the caller's local module call sign, which is **RXRPT2** in the *Received Call Memory.*

The procedure in the next section allows you to copy call signs from the *Received Call Memory* to either the *UrCall* or *Repeater* memory banks, where they can later be used for copying to the *Call Sign Routing Register*.

Copying Calls from the Received Call Memory

If desired, you can copy calls that were automatically recorded in the temporary *Received Call Memory* to either the *UrCall* or *Repeater* Memory banks. This can be useful if you heard a call that you want to save for later use.

Copying Calls from the Received Call Memory on the IC-91AD and IC-92AD:

1. Starting with the DV operating mode selected and a D-STAR repeater frequency being displayed on the screen.
2. Press [**MENU**], which brings up the menu list.
3. Rotate [**DIAL**] to select the **RX CALL S** entry.
4. Press [▶] to access the **RX CALL SIGN** memory bank.
5. Rotate [**DIAL**] or use [▲] / [▼] to select the call record you want to copy from, numbered **RX01** through **RX20**.
6. Press [▶] and then rotate [**DIAL**] to skip over **ALL** and select the call sign you wish to copy, either: **CALLER**, **RXRPT1** or **RXRPT2**.
7. Press [▶] to enter the copy mode.
8. Rotate [**DIAL**] or use [▲] / [▼] to select the **LIST SEL** entry to be able to select the memory channel to be programmed.
9. Press [▶] and then select the memory location you wish to copy into. (Select an empty channel or over write a programmed channel as desired.)
10. Press [▶] to save to the selected location.
11. Press [**MENU**] to exit and return to normal operation.

Copying Calls from the Received Call Memory on the IC-2820:

1. Starting with the DV operating mode selected.
2. Press [**F**] to access the function keys
3. Press [**MENU**] to select the **MENU** screen.
4. Rotate [**MAIN•BAND**] to select **RX CALL SIGN** screen, then press [**MAIN•BAND**].
5. Rotate [**MAIN•BAND**] to select the call record you want to copy from, numbered **RX01** through **RX20**.
6. Press [**MAIN•BAND**] and then rotate [**MAIN•BAND**] to skip over **ALL** and select the call sign you wish to copy from, either: **CALLER**, **RXRPT1** or **RXRPT2**.
7. Press [**MAIN•BAND**] and rotate [**MAIN•BAND**] to select the **LIST SEL** entry to be able to select the memory channel to be programmed.
8. Press [**MAIN•BAND**] and then rotate [**MAIN•BAND**] to select the memory location you wish to copy into. (Select an empty channel or over write a programmed channel as desired.)
9. Press [**MAIN•BAND**] to save to the selected location.
10. Press [**BACK**] 3 times to exit and return to normal operation.

Chapter 5: **DV Short Text Messaging**

You can configure your transceiver to send short messages whenever you transmit. These messages are transmitted along with your voice every time you transmit, and will show up the *RX Message* buffer in other user's radios. Typically up to five messages, each a maximum of 20 characters in length, can be programmed into the radio. Before transmitting you select which of the five you wish to transmit.

The messages are free form and can be used to convey a variety of information. It's too bad they are only 20 characters long, as it can be a challenge to figure out how to fit what you want to say using only 20 characters.

Some common uses:

- Radio configuration
- Your name and location
- Your full name (first and last)
- Special announcements
- Radio settings to use for answering your call
- Reflector you are using
- Your web page or email address

Providing instructions on how users should set their radio to make a return call back to you is an interesting use for short status messaging. I found several Japanese stations using this technique. Their short status messages were formatted similar to this: "Set Ur to /JP1XXX," and I've seen others with the message "Use One Touch Button." Even if you don't use this technique yourself, it's worth remembering if you hear a call and are unsure of how to call back. They might be telling you via short messaging.

Programming DV Short Messages

Programming messages into the radio is relatively straightforward. I like using lower-case characters to help make run together words or abbreviations understandable without having to insert a space. For instance, here is what I have programmed in one of my memories: "Bernie SanDiego NoCo". This comes to exactly 20 characters and would not have fit had I used spaces. NoCo stands for North County and may not be obvious, but SanDiego is quite understandable.

Unfortunately there is a drawback to using lower-case characters. Earlier radios such as the ID-800H, IC-V82 and perhaps others can only display upper-case characters, so a message received on those radios may not display as intended.

Programming and Transmitting Short Messages on the IC-91AD and IC-92AD:

1. Starting with the DV operating mode selected and a D-STAR repeater frequency being displayed on the screen.
2. Press [**MENU**], which brings up the menu list.
3. Rotate [**DIAL**] or use [▲] / [▼] to select **MESSAGE / POS**.
4. Press [▶] to access the **MESSAGE / POSITION** screen.
5. Rotate [**DIAL**] or use [▲] / [▼] to select **TX MESSAGE**.
6. Press [▶] to go to the **TX MESSAGE** editing screen.
7. Rotate [**DIAL**] or use [▲] / [▼] to select one of the five memory channels **Ch01~Ch05** or **OFF**. (Note: **OFF** is used to disable message transmissions.)
8. Press [▶] to start programming your message.
9. Use the keys as described at bottom of the screen to enter your message text.
10. When editing is complete, press [**5/** ↵] to save the message and return to the **TX MESSAGE** screen.
11. Before returning to normal operation, rotate [**DIAL**] or use [▲] / [▼] to select the message channel **Ch01~Ch05** you wish to transmit, or select **OFF** if you wish to disable message transmission.
12. Press [**MENU**] to exit and return to normal operation.

Note: If enabled, the message is transmitted every time you press [**PTT**].

Programming Short Messages on the IC-2820:

1. Starting with the DV operating mode selected.
2. Press [**F**] to access the function keys
3. Press [**MENU**] to select the **MENU** screen.
4. Rotate [**MAIN•BAND**] to select **DV MESSAGE** screen, then press [**MAIN•BAND**].
5. Rotate [**MAIN•BAND**] to select **TX MESSAGE MEMORY.**
6. Press [**MAIN•BAND**] and then rotate [**MAIN•BAND**] to select one of the five TX memory channels **01~05**.
7. Press [**MAIN•BAND**] to start programming your message.
8. Use the key functions at the bottom of screen to enter or edit your message.
   a) Press [**ABC**] to select between lower and upper case
   b) Press [**1/**] to select between numbers and symbols
   c) Rotate [**MAIN•BAND**] to select characters
   d) Use [**<**] and [**>**] keys to move entry position cursor
   e) Press [**CLR**] to clear selected characters
   f) Press [**CLR**]1 sec to clear all characters after the cursor
9. When message is complete, press [**MAIN•BAND**] to save it.
10. To exit the **TX MESSAGE MEMORY** screen and return to normal operation, press [**BACK**] as required.

Transmitting Short Messages on the IC-2820:

1. Starting with the DV operating mode selected.
2. Press [**F**] to access the function keys
3. Press [**MSG**] to select the **MESSAGE** screen.
4. Rotate [**MAIN•BAND**] to select **TX MESSAGE** screen, then press [**MAIN•BAND**].
5. Rotate [**MAIN•BAND**] to select the channel to be transmitted **Ch01 ~ CH05**, or select **OFF** to disable message transmission.
6. Press [**BACK**] as required to return to normal operation.

Note: If enabled, the message is transmitted every time you press [**PTT**].

Reviewing Short Messages

Only the last message received can be viewed. This can present a problem, because it can be overwritten if another message happens to arrive before you have a chance to view the message.

Reviewing Received Short Messages on the IC-91AD and IC-92AD:

1. Starting with the DV operating mode selected and a D-STAR repeater frequency being displayed on the screen.
2. Press [**MENU**], which brings up the menu list.
3. Rotate [**DIAL**] or use [▲] / [▼] to select **MESSAGE / POS**.
4. Press [▶] to access the **MESSAGE / POSITION** screen.
5. Rotate [**DIAL**] or use [▲] / [▼] to select the **RX MESSAGE**.
6. Press [▶] to view the **RX MESSAGE** screen.
7. Rotate [**DIAL**] or use [▲] / [▼] to toggle between viewing the received **MESSAGE** or the **CALLER**'s call sign.
8. Press [**5/ ↵**] to return to the **RX MESSAGE** screen.
9. Press [**MENU**] to exit and return to normal operation.

Reviewing Received Status Messages on the IC-2820:

1. Starting with the DV operating mode selected.
2. Press [**F**] to access the function keys
3. Press [**MENU**] to select the **MENU** screen.
4. Rotate [**MAIN•BAND**] to select **DV MESSAGE** screen, then press [**MAIN•BAND**].
5. Rotate [**MAIN•BAND**] to select **RX MESSAGE**.
6. Press [**MAIN•BAND**] to view the **RX MESSAGE** screen.
7. Press [**BACK**] 3 times to exit and return to normal operation.

Chapter 6: **Internet Resources**

Amateur radio is fortunate to have enterprising programmers whom have authored a number of innovative and useful D-STAR web page programs. Lets examine a couple of the most popular ones: *D-STAR Calculator* and *jFindu.*

D-STAR Routing and Linking Calculator

To assist amateurs with programming the call routing parameters for their radios, Ed Woodrich, WA4YIH has developed an easy to use program, *D-STAR Calculator*. The program operates on-line via the Internet and uses pull-down menus to make repeater and routing selections.

You start out by selecting the local repeater and module that you are using to get to the gateway. Next you select the D-STAR call routing mode that will be used to route the call. Finally the remote repeater module you want your signal to come out of is selected. After all this information has been selected, the calculator displays the settings you need to program into your radio for making the call.

This program really simplifies finding repeaters and programming the *Call Sign Routing Register*; anybody can do it!

Operating the D-STAR Calculator Program

This program contains a full list of repeaters and reflectors that are available for use worldwide. US repeaters are listed by state and city. If you want to make a call to a foreign country, a pull-down list makes it easy to determine which repeaters are available by country, province and city.

In the sample screen shot below, after I selected my local repeater using the *Source Repeater* pull-down menu, the program displayed a list of repeater modules that are available on the selected repeater. In this case there was only one *Source Module* displayed, so I selected it.

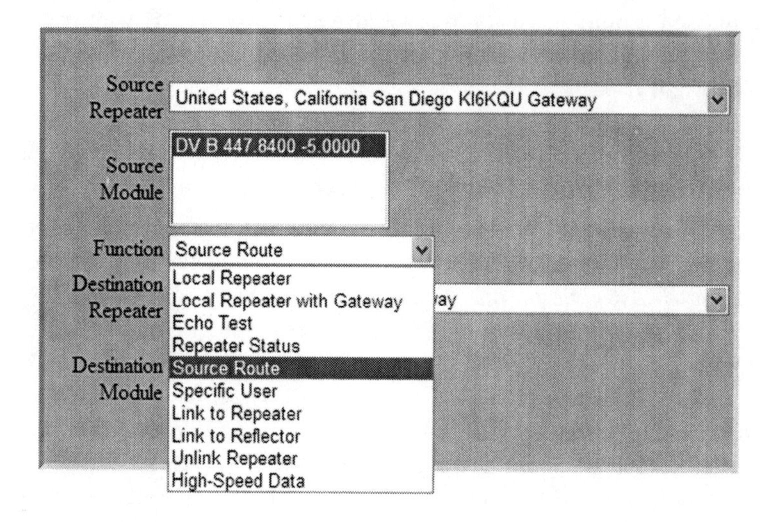

As shown above, I then selected the D-STAR routing / linking *Function* that I wanted to use from a list of available functions. The list includes all the routing and linking modes that were discussed in Chapters 1 and 2 plus one more, *High-Speed Data*. In this example I selected *Source Route* as the mode of routing I wanted to use.

Continuing with making selections, as in the screen shot below, I selected Canada, Calgary VE6WRN, as my *Destination Repeater* and module B as my *Destination Module*, which are the last pieces of information needed by the calculator.

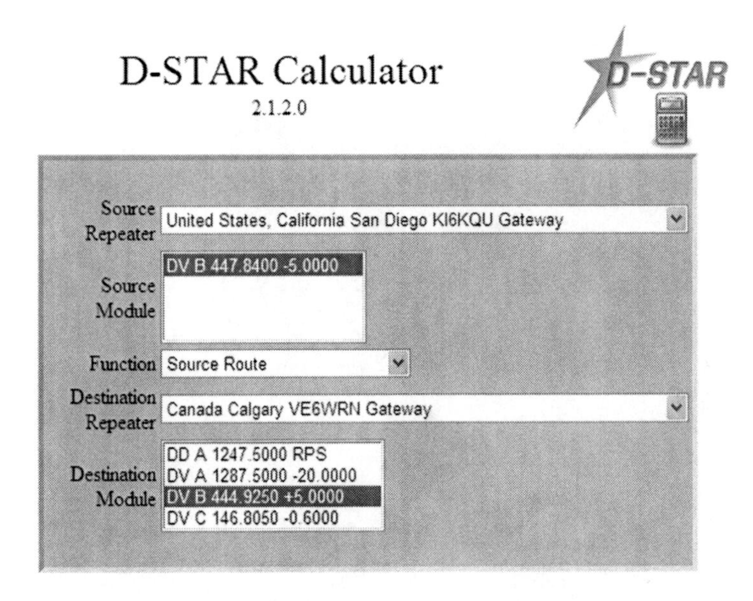

After selecting the *Destination Module*, the calculator came up with the results shown in the following screen shot. The shaded box contains the information needed to program your radio's *Call Sign Routing Register* fields: **UrCall**, **RPT1** and **RPT2**. If you need it, the frequency and offset for your local repeater are also shown.

Below the shaded box, the program displays a call routing diagram. At the sides and bottom are sample scripts suggesting how to identify when making the call. Besides giving their location and the repeater module they are using, the scripts show the two hams exchanging repeater frequency and offset information. Typically when operating on D-STAR, exchanging repeater frequencies is unnecessary.

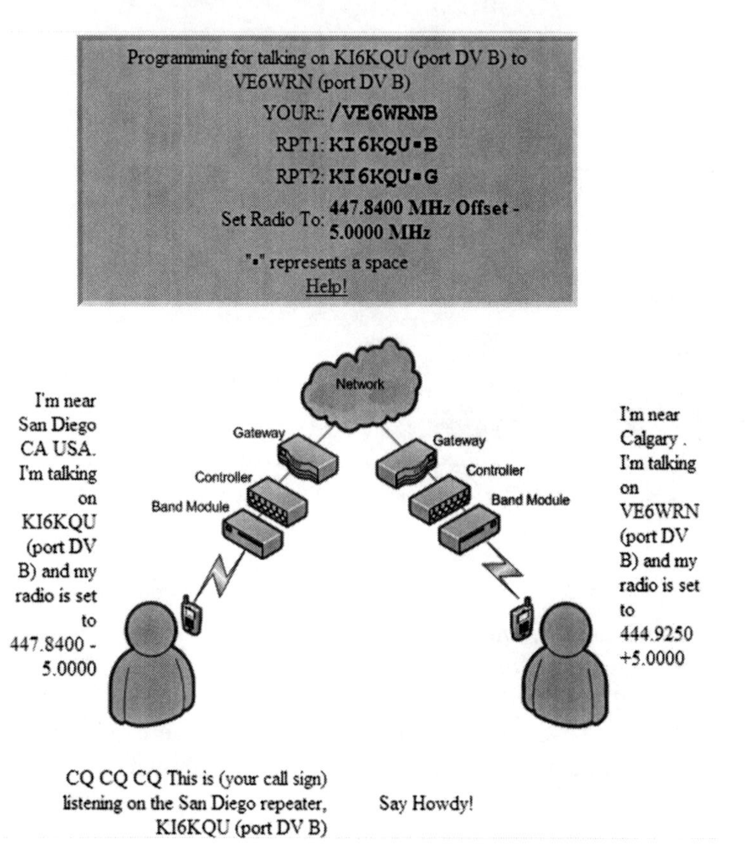

As you can see, the program is simple to use and by making a few list driven selections you are provided with all the parameters necessary for programming your radio to make the call.

D-STAR Calculator can be found at the following URL:
http://www.dstarinfo.com/Calculator/

jFindu Repeater Locator and Last Heard Lists

Another useful function is to be able to find out when, and on what repeater a station was last heard. By clicking on the **Last Heard Lists** link found at: **http://www.jfindu.net** home page you can find out. In the example below, I have input WD6FZA and clicked the "Include D-STAR last heard?" box.

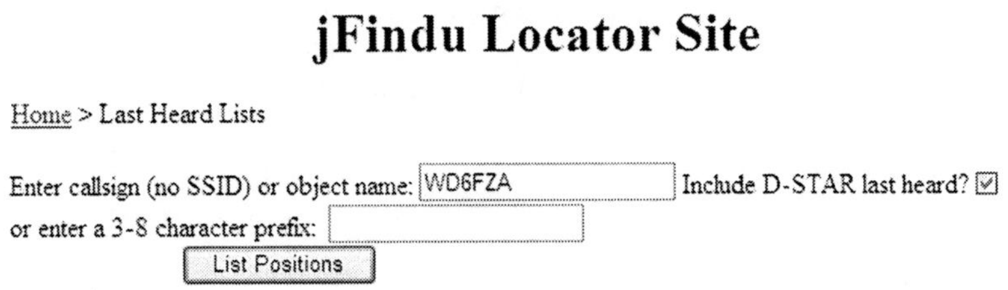

jFindu Locator Site

Home > Last Heard Lists

Enter callsign (no SSID) or object name: WD6FZA Include D-STAR last heard? ☑

or enter a 3-8 character prefix:

List Positions

The result is displayed as shown below. If the person has been using APRS, a status line for that mode is also displayed.

WD6FZA Last Heard on DSTAR

Station	Last Heard	Repeater
WD6FZA	3h47m43s	KI6KQU B(DV)

Another useful *jFindu* function is to find out where a repeater is located, what modules it has and who has been using it recently. This information as well as recent D-STAR D-PRS / APRS activity can be determined by accessing **http://www.jfindu.net** and clicking on the **D-STAR Activity** link and when the next page displays, clicking on **D-STAR Repeaters.**

The **D-STAR Repeaters** link brings up a page showing a world map, which will gradually populate itself with the call signs of all D-STAR repeaters. These call signs are geographically located on a satellite-view world map. Like many on-line maps, you can zoom in and scroll around to see repeater installation density in different parts of the world. If you zoom in far enough you can see where repeaters are located near a city or on mountain ranges.

More importantly however, is the matrix of active repeater D-STAR call signs that is below the map. Click on a repeater call sign of interest to see detailed information about that repeater, as in this screen shot for the VE6WRN repeater located near Calgary Canada.

Locate VE6WRN Repeaters

A	Range: 20nm 1.2 Voice 1287.5000 -20 Mhz
AD	Range: 15nm 1.2 Data 1247.500 Mhz
B	Range: 30nm 440 Voice 444.925 +5.00 Mhz
C	Range: 30nm 2m Voice 146.805 -0.600 Mhz

Recently Heard DV Stations

Station	LastHeard	Repeater
VY0AB	6d22h56m4s	VE6WRN A
VE6WRN	17h31m34s	VE6WRN B
VE6BGZ	4d4h35m12s	VE6WRN B
G4HCI	21d1h50m4s	VE6WRN B
VE6BGZ M	19h13m22s	VE6WRN C
VE6CPK	19h13m42s	VE6WRN C
VE6AMC	22h30m38s	VE6WRN C
VE6BGZ P	4d1h48m13s	VE6WRN C
VE6BNF	5d16h31m	VE6WRN C
VE6DJJ	6d3h50m52s	VE6WRN C
VE6CPT	17d12h13m28s	VE6WRN C

The screen above shows the available modules on this repeater, including their frequency and offset. Below that, sorted by module, is a list of stations last heard, including how long it's been since they were heard.

If you happen to pull up this screen for your own local repeater, you will be able to track recent traffic as you are hearing it on your radio, after a delay of a few moments. For instance, if you make a transmission, your own call sign will eventually appear in the list.

Chapter 7: **Radio Programming Software**

If you are going to be programming a lot of repeater frequencies and D-STAR access call signs into your radio, you should definitely consider purchasing programming software for your radio. Not only is it much easier to do, you might be able to import programming files from others that can significantly simplify setting up your radio.

Another plus for programming software is that once you have your radio setup, a copy of your radio's frequency memories and setup menus can be saved to your computer. This can be a lifesaver should you need to reload your radio for any reason.

Besides the software itself, an interface cable for connecting between the radio and your computer is required. Most of the time, a standard RS-232 PC serial port is used on the computer side. However, on the radio side, the connector is often unique to the radio. Frequently the required cable comes with the purchase of the programming software.

Icom's Programming Software

Icom provides programming software (frequently called cloning software) for their radios. The following list identifies the software packages used with Icom's D-STAR radios. Most of these come with a cable, but some do not.

- IC-91AD RS-91 software
- IC-92AD RS-92
- IC-V82 CS-V82
- IC-2820 CS-2820
- IC-2200 CS-2200
- IC-D800 CS-D800

RT System's Programming Software

RT-Systems also provides radio-programming software for radios of all types. Their software kits generally come with the required cable. Here are the RT part numbers for the software kits used with Icom's D-STAR radios

- IC-91AD WCS-91
- IC-92AD WCS-92
- IC-2820 WCS-2820
- IC-2200 WCS-2200
- IC-D800 WCS-D800

I have not used any of RT System's software for D-STAR radios, but have used their programs for several conventional transceivers. The software I've used was reliable and did the job. Their software is geared towards programming your radio and does not include the "virtual radio" features discussed below.

Icom's RS-91 and RS-92 Programming Software

Most all programs provide similar programming functionality; some of them go a step beyond. Here we will review the features of Icom's RS-92 (Remote Control) software application as an example of what can be done with one of the more enhanced programming packages.

Unlike most other programming software packages, the RS-91 and RS-92 software packages have the capability of actually controlling the radio via a "virtual radio" looking screen. Using this "virtual radio" is quiet similar to using a remote front panel, the connected radio responds just as if you were pushing directly on the radio's front panel buttons. In general, radio-programming software packages do not have this "virtual radio" capability; it is limited to Icom's RS, Remote Control series of software. Icom's CS series Cloning Software is more typical of what is usually available for programming radios.

When the RS-92 program is started, it comes up with the "virtual radio" screen. The first thing you need to do is click on the **Option** pull-down menu and select the PC COM port that you will be using. The program will read the radio's current settings, which takes a second or two, then the "virtual screen" will display the same information that is being displayed on the radio's screen. While the computer is reading the balance of your radio's settings (called synchronizing), the () button near the top of the display will be red; synchronization is complete when it turns black.

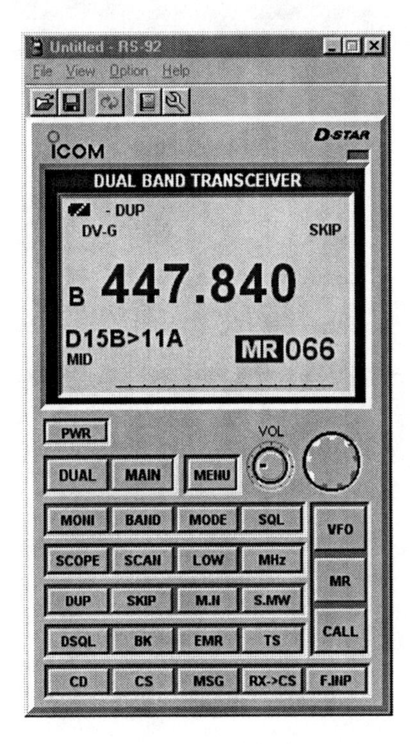

To examine or edit the radio's memory, wait for the synchronization to complete (() button turns black) and click on the **View** pull-down and select **Edit Memory Channel** and the **Memory CH** screen pops up. At the left side of the screen, a series of folders and sub-folders are displayed in typical Windows format. These folders contain all of the memories, setup menus and other parameters that can be programmed on the radio.

The frequency memory channels are partitioned into two folders **A Band** and **B Band**, within these two folders the memory channels are displayed in 100 channel increments. Only the **B Band** can be used for DV mode operation. The two screenshots below shows a selection of memories from the first group of memories in the **B Band / ALL** folder. These screenshots are segments of a wider display as seen on a PC.

Memory CH

		Frequency						
CH	Select	Freq	DUP	Offset Freq	TS	Mode	Name	Skip
13		445.86000	-DUP	5.00000	12.5k	DV	D-10B	
14		145.58500	-DUP	0.60000	15k	DV	D-10C	
15		446.98000	-DUP	5.00000	12.5k	DV	D-11	
16		447.84000	-DUP	5.00000	12.5k	DV	D-15	
17		448.70000	-DUP	5.00000	12.5k	DV	D-16	Skip
18		447.20000	-DUP	5.00000	12.5k	DV	D1 CQ	Skip
19		447.20000	-DUP	5.00000	12.5k	DV	D1 ECHO	Skip
20		447.20000	-DUP	5.00000	12.5k	DV	D1 > D4	Skip
21		447.20000	-DUP	5.00000	12.5k	DV	D1 >D10B	Skip
22		447.20000	-DUP	5.00000	12.5k	DV	D1> D10C	Skip
23		447.20000	-DUP	5.00000	12.5k	DV	D1 > D11	Skip
24		447.20000	-DUP	5.00000	12.5k	DV	D1> D15B	Skip
25		447.20000	-DUP	5.00000	12.5k	DV	D1 > D16	Skip

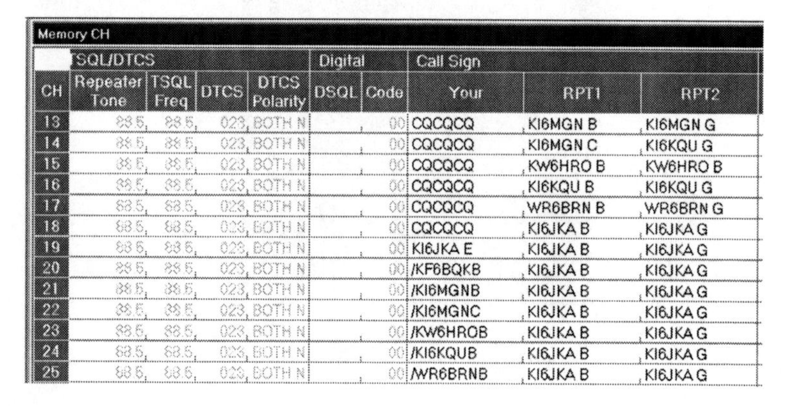

Memory CH

	TSQL/DTCS				Digital		Call Sign		
CH	Repeater Tone	TSQL Freq	DTCS	DTCS Polarity	DSQL	Code	Your	RPT1	RPT2
13	88.5	88.5	023	BOTH N		00	CQCQCQ	KI6MGN B	KI6MGN G
14	88.5	88.5	023	BOTH N		00	CQCQCQ	KI6MGN C	KI6KQU G
15	88.5	88.5	023	BOTH N		00	CQCQCQ	KW6HRO B	KW6HRO B
16	88.5	88.5	023	BOTH N		00	CQCQCQ	KI6KQU B	KI6KQU G
17	88.5	88.5	023	BOTH N		00	CQCQCQ	WR6BRN B	WR6BRN G
18	88.5	88.5	023	BOTH N		00	CQCQCQ	KI6JKA B	KI6JKA G
19	88.5	88.5	023	BOTH N		00	KI6JKA E	KI6JKA B	KI6JKA G
20	88.5	88.5	023	BOTH N		00	/KF6BQKB	KI6JKA B	KI6JKA G
21	88.5	88.5	023	BOTH N		00	/KI6MGNB	KI6JKA B	KI6JKA G
22	88.5	88.5	023	BOTH N		00	/KI6MGNC	KI6JKA B	KI6JKA G
23	88.5	88.5	023	BOTH N		00	/KW6HROB	KI6JKA B	KI6JKA G
24	88.5	88.5	023	BOTH N		00	/KI6KQUB	KI6JKA B	KI6JKA G
25	88.5	88.5	023	BOTH N		00	/WR6BRNB	KI6JKA B	KI6JKA G

The above screenshots show the contents for memory channels 13 through 25, which have been programmed for D-STAR operation. The upper one shows the repeater frequency settings, tuning step size, operating mode and the name / label given to this memory channel. The lower screenshot shows the D-STAR call sign routing parameters set for each of these memory channels.

Editing a memory channel, is a simple matter of clicking in one of the cells, and making whatever changes you want, similar to making entries on a spreadsheet. Some of the cells have a pull-down selection menu, which is displayed if you hit Enter on the keyboard. Pull-down selections are used for parameters that have several standard settings, such as: Duplex, Tuning Step Size, Operating Mode, etc.

You can select setup menus or other memories to be viewed or edited by either clicking on the file folders shown at the left of the screen or by using the **View** pull-down selection menu. The following screen shows the **DV Set Mode** menu settings, which are one of the sub groups found under the **Digital Setting** group of menu settings.

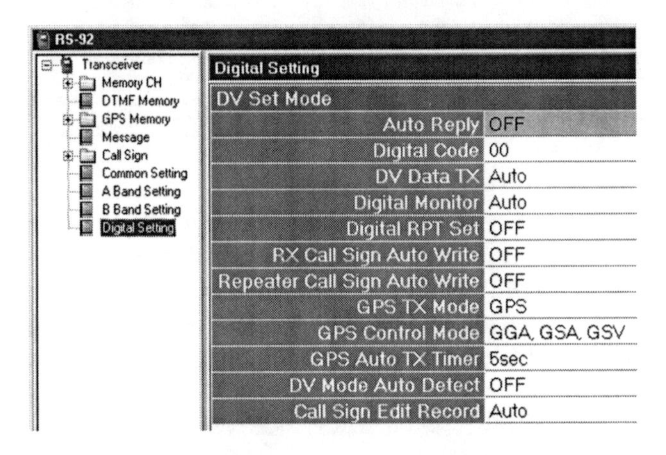

Like most PC applications, the **File** pull-down on the "virtual radio" window allows you to save your settings to the PC and download files containing memories and configuration settings to the radio.

D-STAR Operation Using the RS-92 Software

One of the great benefits of using the RS-91 and RS-92 software is that you can actually operate the radio via screens on your PC. This can be a whole new D-STAR operating experience.

While monitoring activity on a repeater, received call signs and the caller's D-STAR call sign routing parameters are dynamically displayed in the **Received call record** as calls are received. These entries are the same ones that are found in the *Received Call Sign History* found on the radio, but without having to go into the radio's menu system to see them. Because the **Received Call Record** displayed on the PC shows all of a caller's data on a single line, with multiple lines showing prior received calls, its much more convenient than accessing the same data via the radio's keypad and small screen.

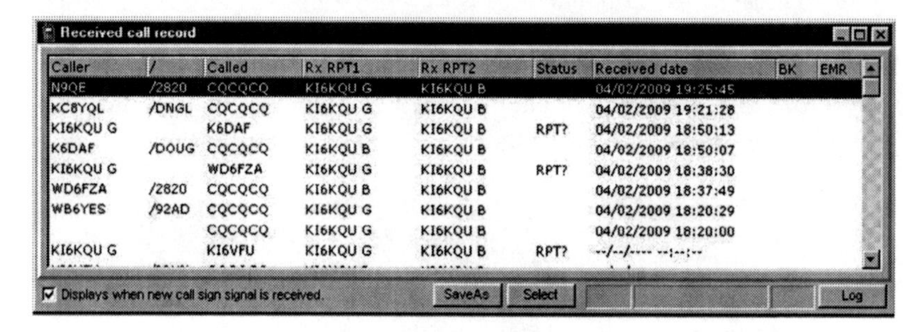

Likewise, as seen below, the short messages of stations being monitored are displayed as calls are received. A big advantage of the RS-92's **Message Reception and Transmission** window is that while the <u>radio</u> is limited to saving only the last received message, this screen displays a list of prior calls in addition to the most current.

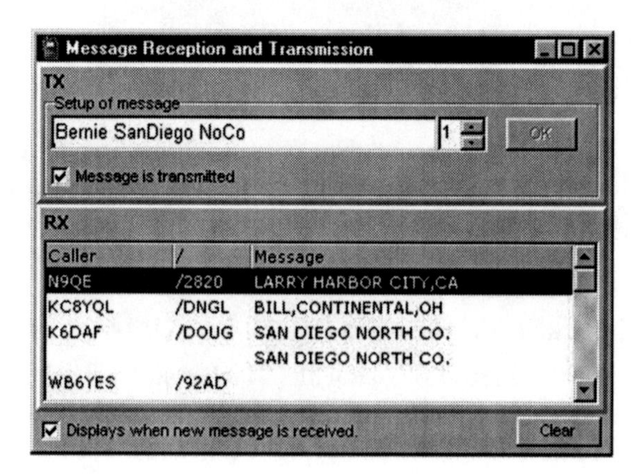

Received messages are displayed in the lower half of the window. The upper section lets you format and select short messages of your own. You can have up to five of these, and the one currently showing is the one that will be transmitted.

In the lower left corner of these two windows is a box that can be checked to have the window pop-up whenever a new record is received. Depending upon what you are trying to do, having these windows popping up can be annoying. If so, just unclick the **Displays when new message is received** box.

Another useful screen that facilitates D-STAR operation is the **Select Call Sign** window. This handy screen lets you conveniently configure the *Call Sign Routing Registers*.

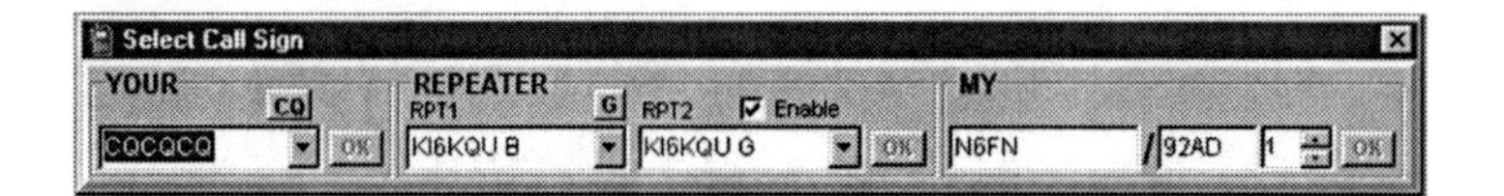

When operating from a station where you can use a PC, the windows described above can all be placed on the PC's desktop and monitored while operating. It's quite informative to see the data displayed as calls are received, especially when contacting hams in foreign countries.

Chapter 8: **DV Mode Slow-speed Data**

Icom's D-STAR compatible VHF and UHF transceivers come factory ready for 1200 bps data communication. All you need is an RS-232 interface cable for connection to a PC, some data communication software and you're in business. In general you can use the same cable that is used to program your radio.

Cables for different model transceivers are available from Icom, or you can make your own. Information for fabricating your own data cables can be found at **http://epaares.org/dstar/icom_cables.htm**.

DV mode digital data can be sent simultaneously with voice transmission. As explained in Chapter 1, the basic D-STAR digital packet structure has space reserved for data and voice in the same packet. DV mode transmission consists of a continuous stream of digital packets for as long as you are transmitting. Even when you are not transmitting data, the packet structure still includes space for the data. Voice and data transmission on the same signal are inseparable, even though one or the other are may not be used. As a result, voice communication is not adversely impacted while data is being sent at the same time. Conversely, data throughput does not increase when not transmitting voice.

Data transmission through the D-STAR system is software and protocol independent. To software applications, the PC's data path to the originating transceiver, through the repeater's gateway to the Internet and on to similar equipment at the other end, appears just like a cable. Even though the data path is complex, software running on a PC is not involved with any of the packet formation or routing issues. All of that is handled by the transceiver and the gateway. The PC's software only sees a stream of data bytes, stripped of all packet header and routing information.

Since the entire data transport pathway, as complex as it is, appears no different than a cable to software applications, existing terminal emulation or data transport programs can be used without modification. Familiar programs such as *Hyper Terminal*, found on most PC's, and *ProComm*, a long time favorite for modem communication, can readily be used. The catch is that the user at the other end has to either use the same program, or one that is at least compatible with the data that is being sent by the originating program.

D-STAR Oriented Data Communication Software

Of course, hams are never satisfied with just making do with what is readily available. Several have developed D-STAR data communication software applications that seek to improve performance and capability beyond programs originally developed for old-style modem communications. While there are a number of D-STAR data communication programs out there, the two most popular seem to be *d*Chat* and *D-RATS.*

*d*Chat* (D-STAR Chat) is a simple to use Windows based keyboard-to-keyboard communication application. Keyboard-to-keyboard communication on D-STAR is somewhat similar to using RTTY or PSK31, except text data is transmitted a line at a time. This is probably the program to use if you want to experiment with keyboard-to-keyboard communication with a friend or two.

D-RATS on the other hand, while still capable of keyboard-to-keyboard communication, sports a much more extensive list of features such as: file transfers, GPS position reporting and mapping, message filtering, email reception and forwarding and even a method for creating and using forms. Several premade forms can be used for sending eMail, radiograms and other purposes. Also included are several sample forms for supporting EMCOMM incident communications. Existing forms can be edited to suit your purposes, or you can create your own. *D-RATS* is a multi-platform application that can run on Windows, Linux/UNIX and MacOS X operating systems.

Radio / PC Configuration for Low-speed data Operation

Setup for low-speed data operation is relatively simple. Other than configuring the PC communication software application that will be used, there are three things to be concerned about on the radio:

- Serial port settings
- Selection of Automatic or PTT triggered data transmission
- Disabling GPS transmission

Configuring Serial Ports

The serial port data rate for some Icom transceivers is fixed; for others it can be set via a setup menu. This chart identifies serial port data rate setting capability by radio model.

Radio Model	Fixed Data Rate	Use Menu to Set
IC-2820	9600 bps	na
IC-92AD	38.4 kbps	na
IC-91AD	38.4 kbps	na
IC-V/U82	na	9600 bps
IC-2200H	na	9600 bps
ID-800H	na	9600 bps

The PC's serial port must be set to match the radio's serial port parameters as follows:

- Baud Rate Set as per the above chart
- Data 8 bit
- Parity none
- Stop bits 1 bit
- Flow control Xon / Xoff

Typically the PC's serial port parameters are set using the PC's communication software application.

Automatic / PTT Data Transmission Selection

On all D-STAR compatible Icom radios, except for the IC-V/U82 there is a menu for selecting either Automatic or PTT triggered data transmission. PTT is typically the radio's default condition, and limits data transmission to when the microphone's PTT key is pressed. When set to Auto, the radio will start transmitting data whenever it shows up from the PC's communication software application. (The IC-V/U82 is fixed to operate in the Auto mode.)

Typically when transmitting data, the radio should be set to automatically initiate transmission whenever data is received from the PC. But you need to keep in mind that when operating over a repeater, especially if you are linked to a reflector, that the PC will send data whenever it has it. If sending GPS position data or operating other "beacon" modes this can be a problem because it can collide with other user's transmissions. In the case of sending GPS position data, the PTT menu option should be selected to prevent continuous "pinging" of the repeater.

The procedures for selecting either Auto or PTT data transmission for the IC-91AD, IC-92AD and the IC-2820 are shown below. Consult the Icom user manual for the IC-2200H and the ID-800H.

Automatic / PTT Selection on the IC-91AD and IC-92AD:

1. Starting with the DV operating mode selected and a D-STAR repeater frequency being displayed on the screen.
2. Press [**MENU**], which brings up the menu list.
3. Rotate [**DIAL**] or use [▲] / [▼] to select **DV SET MODE**.
4. Press [▶] to the **DV SET MODE** menu.
5. Rotate [**DIAL**] or use [▲] / [▼] to select **DV DATA TX**.
6. Press [▶] to the **DV DATA TX** option setting screen.
7. Rotate [**DIAL**] or use [▲] / [▼] to select either **PTT** or **AUTO**.
8. Press [**MENU**] to exit and return to normal operation.

Automatic / PTT Selection on the IC-2820:

1. Starting with the DV operating mode selected.
2. Press [**F**] to access the function keys
3. Press [**MENU**] to select the **MENU** screen.
4. Rotate [**MAIN•BAND**] to select **DV SET MODE** screen, then press [**MAIN•BAND**].
5. Rotate [**MAIN•BAND**] to select **DV DATA TX**, then press [**MAIN•BAND**].
6. Rotate [**MAIN•BAND**] to select either **PTT** or **AUTO**.
7. Press [**BACK**] 3 times to exit and return to normal operation.

Disabling GPS Mode Transmission

To be able to transmit slow-speed data with the IC-91AD, IC-92AD and IC-2820 transceivers its first necessary to turn off GPS mode operation, otherwise the radio will not transmit. For other radios, consult the Icom user manual to see if GPS needs to be turned off.

Disabling GPS Transmission on the IC-91AD and IC-92AD:

1. Starting with the DV operating mode selected and a D-STAR repeater frequency being displayed on the screen.
2. Press [**MENU**], which brings up the menu list.
3. Rotate [**DIAL**] or use [▲] / [▼] to select **DV SET MODE**.
4. Press [▶] to the **DV SET MODE** menu.
5. Rotate [**DIAL**] or use [▲] / [▼] to select **GPS TX MODE**.
   (On the IC-91AD select **GPS MODE**)
6. Press [▶] to the **GPS TX Mode** option setting screen.
7. Rotate [**DIAL**] or use [▲] / [▼] to select **DISABLE** or **OFF**.
8. Press [**MENU**] to exit and return to normal operation.

Disabling GPS Transmission on the IC-2820:

1. Starting with the DV operating mode selected.
2. Press [**F**] to access the function keys
3. Press [**MENU**] to select the **MENU** screen.
4. Rotate [**MAIN•BAND**] to select **DV GPS** screen, then press [**MAIN•BAND**].
5. Rotate [**MAIN•BAND**] to select **GPS AUTO TX**, then press [**MAIN•BAND**].
6. Rotate [**MAIN•BAND**] to select **OFF**.
7. Press [**BACK**] 3 times to exit and return to normal operation.

d*Chat Application Installation and Setup

*d*Chat*, written and maintained by Brian Roode, NJ6N can be downloaded from his web page **http://nj6n.com/dstar/dstar_chat.html**

Installation is simple and straightforward following the instructions found on the web page. I downloaded the program to my computer by clicking on the link in step 3 of his instructions, which seemed like the simpler of several options presented. The program downloaded is a .zip file that you can extract to some convenient location. It installs much like any other Windows application and places a **dChat2** program entry your computer's **Start, All Programs** list.

Start the program by clicking on **dChat2** in the **All Programs** list. When the **dChat nj6n** window comes up, click on **Settings** at the top of the screen to display the **Settings** window. Here you can enter your **Call Sign** and set the **Communications Port Settings** by selecting the COM port that is connected to your radio and it's baud rate. Set the baud rate to one that is supported by your radio.

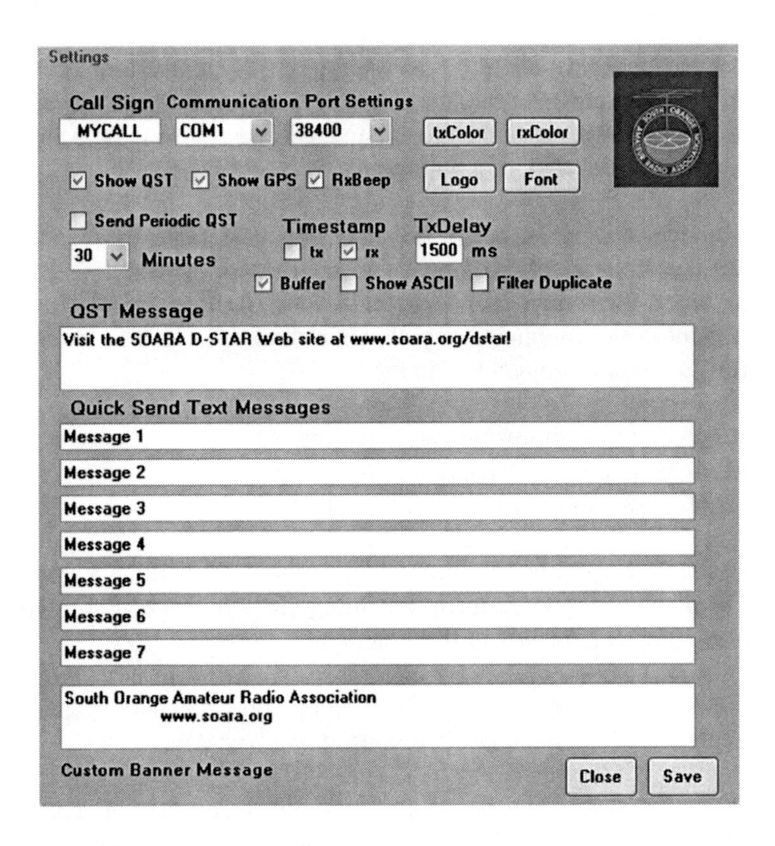

The **txColor** and **rxColor** buttons are used to set the color for text that appears in the text box on the main screen. I selected red for transmit and blue for receive. The **Font** button selects the font style for text that is displayed.

The **Settings** screen contains two areas for pre-formatting messages that you want to send. In the **QST Message** area, text that is to be broadcast on a periodic basis can be programmed. If you want to broadcast QST messages, set the **Minutes** pull-down to a suitable time between broadcasts. In the screen above, its been set to once every 30 minutes, and then check the **Send Periodic QST** box. Since this will periodically ping the repeater, you want to be careful about using this.

The **Quick Send Text Messages** area allows you to program up to seven different messages that can later be easily recalled and transmitted. These are intended for messages that you might use on a regular basis. These could be for sending a CQ, instructions on how to reply to your call, station identification, equipment setup and so on. Once programmed, these messages can easily be transmitted from the main screen at the click of a button.

The **Custom Banner Message** box lets you customize the banner / header that appears at the top of the main **dChat** screen. Here you could enter your own call sign, club identification, special event identification or anything else you might think of. This is "window dressing" and does not get transmitted.

The **Logo** button is more "window dressing" and lets you select a .jpg image of your own to replace the logo displayed in the upper right corner of the **Settings** and main **dChat** screens.

When the **Show GPS** box is checked, incoming GPS / DPRS messages will be displayed. If too many incoming GPS messages get to be a problem, uncheck the box to filter them out.

The **Filter Duplicate** box lets you filter out received duplicate messages based upon the last two lines of received data.

Normally the **Buffer** box is left checked, but if you need to see non line-terminated data (text run together without line breaks), such as might be received from D-RATS, uncheck this box.

Normally the **Show ASCII** box is left unchecked. If checked the ASCII code representing the character will be displayed in parenthesis following each character. This might be used for troubleshooting or other purposes.

The **TxDelay** box is used to set the maximum random delay that is used between each line that is sent. A random delay is inserted to help avoid collisions (simultaneous transmissions with other stations). Normally the default setting of 1500 ms should be fine.

When you are finished with the Settings screen, be sure to click the **Save** button at the bottom to save you work. To go back to the main **dChat** screen, click on the **Settings** button.

d*Chat Program Operation

Before running the program, make sure the following items have been set on your radio. If necessary, refer to the above sections for information on setting these parameters.

- If required for your radio, set the radio's serial port settings.
- Select Automatic data transmission
- If required for your radio, disable GPS transmission

Once installed and setup, *d*Chat* is relatively simple to operate. When started, the program displays the following screen.

First, making sure your radio is connected and tuned to a D-STAR repeater or DV mode simplex frequency, clicking on the **Connect** button should result in a **COMx opened** port status being displayed just to the left of the Recall button. (**COM3 opened** is shown above)

Transmitting consists of sending some text by either typing in the **Text to Send** box or by clicking on one of the messages in the **Quick Send** box, which copies the message to the **Text to Send** box.

When the data in the **Text to Send** box is ready to be sent, click on the **Send** button or hit the Enter key on your computer's keyboard. Data is transmitted a line at a time.

If the text you sent is immediately echoed back, and the **NO DIALTONE** error message is displayed, it was not actually transmitted and you may not have selected the correct COM port or baud rate on the PC.

If you wish to re-send the last line sent, press the **Recall** button.

If your text box starts to get cluttered, it can be cleared by clicking on the **Clear** button.

Pre-formatted small messages can also be broadcast by selecting **Send Bulletin**, under the **File** pull-down menu. This feature cannot be used for file transfers. Instead, it is a way of pre-formatting small multi-line messages for transmission. Messages sent this way show up in other users text window, just like any other messages.

D-RATS Application Installation and Setup

D-RATS, written and maintained by Dan Smith, KK7DS has more features and consequently is a bit more complex to use than *d*Chat*. Dan actively encourages user input, periodically providing new releases implementing changes requested by *D-RATS* users. The program can be downloaded from his web page **http://d-rats.com**

Installation instructions and downloads are provided for the Windows, MacOS X and Linux operating systems. Of the two options provided for Windows installation, I took the simpler option and downloaded and installed the binary release version of the program, which has everything needed for the installation packed into a single zip file. Installation is simple and straightforward following the instructions found on the web page. Unzip the download to a convenient location.

The program does not install itself into your computer's **Start** > **All Programs** list, instead it's started by accessing the folder you placed the program in and clicking on the *d-rats.exe* file. For convenience you can save a short cut to the desktop, or drag the *d-rats.exe* file to your **Start** > **All Programs** list.

To setup the program, start the application and on the **D-RATS** main screen select **File** > **Main Settings**, which displays the **Config** screen.

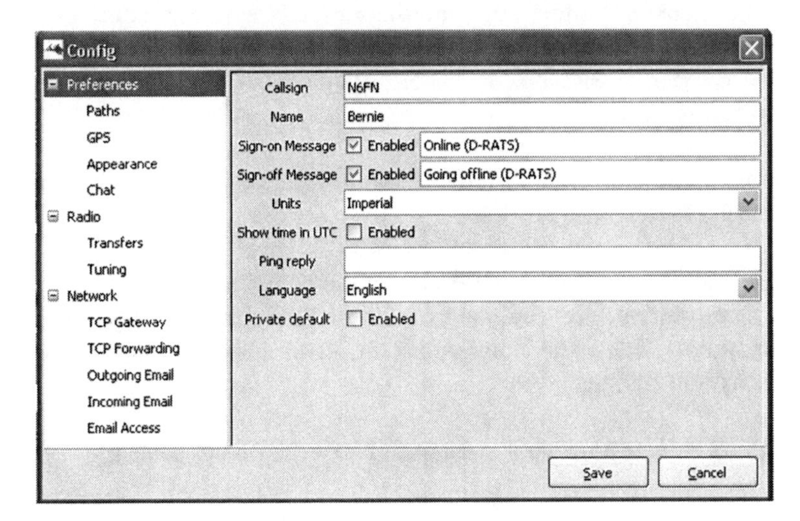

Under the **Preferences** section, enter your **Call Sign** and **Name**. When the **Sign-on** and **Sign-off Message** boxes are checked, short messages are automatically transmitted, indicating whenever you open and close the *D-RATS* program. Uncheck the boxes if you don't want these messages transmitted every time you start the program.

On the same screen you will also see the **Ping reply** box. When another D-RATS user issues a **Ping Station** command, either to a specific station or all stations via CQCQCQ, any online D-RATS stations will automatically respond with an answer. This is where you can program your own reply to a ping. Typically this might be your call, location, a greeting, equipment setup or anything else you desire.

If you intend to perform file transfers, under the **Preferences** section click on **Paths** to set a file folder to which incoming files will be downloaded. The default **Download Directory** setting places them on the desktop. You will notice that you can also select a folder for where maps will be stored.

Also under the **Preferences** section, by clicking on **GPS** you can either enable an attached GPS unit, or set a fixed **Latitude**, **Longitude** and **Altitude**. If you don't attach a GPS, providing you are connected to the Internet, you can easily set the Latitude and Longitude by clicking on the **Lookup** button and entering an address, zip code or an intersection. If connecting an **External GPS** to your computer, check the **Use External GPS** box to enable it and select the COM port and baud rate to be used.

Initially, you can use the default settings for **Appearance** and **Chat** under the **Preferences** section

Next, click on the **Radio** section to set the computer's **Serial Port** and baud rate setting. Set the baud rate to be the same as being used on your radio. The other items under the **Radio** heading can be left at their default settings.

Click on **Save** to save your settings and exit the **Config** screen.

D-RATS Program Operation

Before running the program, make sure the following items have been set on your radio. If necessary, refer to the sections earlier in this chapter for information on setting these parameters.

- If required for your radio, set the radio's serial port settings.
- Select Automatic data transmission
- If required for your radio, disable GPS transmission

You are now ready to receive and transmit messages. Anything typed in the send box will be sent as an entire line, either by clicking on the **Send** button or by pressing the **Enter** key on your computer. Unlike PSK31 or other programs, text is not sent character-by-character as you type; it's sent a line-at-a-time. Sending text a line-at-a-time improves throughput by dramatically reducing the number of packets sent over the D-STAR network.

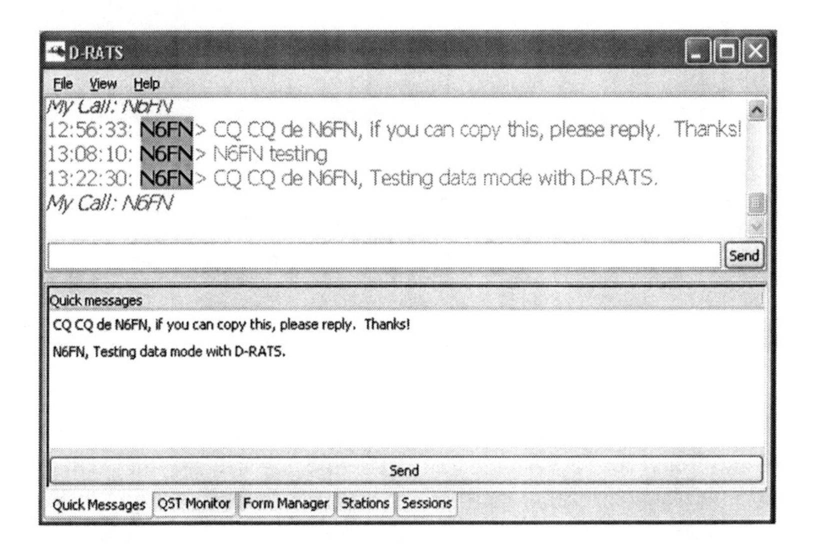

If you wish to clear the upper messages window, click on the **View** pull-down menu and select **Clear**. Even after clearing the upper message window, you can retrieve older data by clicking on the **View** pull-down menu and selecting **Log for this tab**.

If you wish to pre-format short messages that you might use on a regular basis, select **File**, **Quick Messages**, which displays the **Quick Messages** editing screen. Here you can enter, edit and re-arrange messages for later use. Quick messages can be transmitted by selecting the **Quick Message** tab on the main **D-RATS'** screen and then double clicking on the message you want to send.

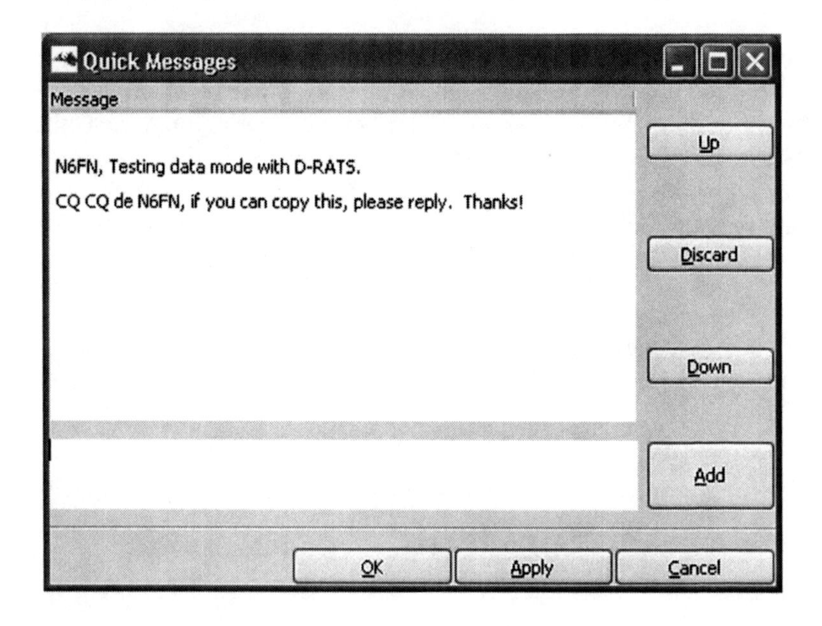

To determine if any other stations are online, you can send a ping command, and any connected D-RATS stations will respond with the reply that was set in their station's configuration setup **Ping reply** box. The **Ping Station** button is found under the **File** pull-down menu. You can ping specific stations that have been recently been heard by using the pull-down box, or by entering the stations call directly into the box. Enter CQCQCQ in the box to receive replies from any stations that are online.

Configuring and Sending D-RATS QST Messages

QST messages can automatically be transmitted on a periodic basis. To pre-format QST Messages, click on the **File** pull-down menu and select **Auto QST Settings**, which brings up a screen for creating the message and selecting the time-to-delay between transmissions.

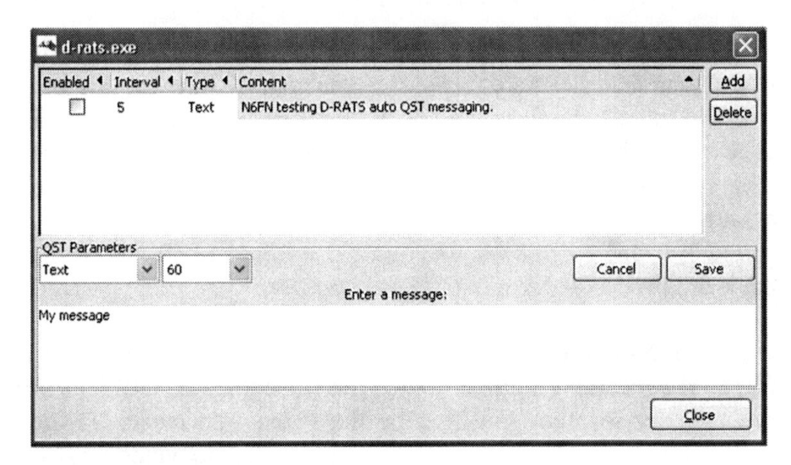

Checking the **Enabled** box selects the QST Message to be transmitted. Upon saving your settings and closing the screen, providing **File, QSTs Enabled** has been selected; the message will be transmitted at the selected time intervals. While the QST Message is being displayed in the **QST Monitor** window, you will see it counting down the transmission timer. If you wish, the message can be immediately sent by double clicking it, handy for bypassing the full time period when testing.

Transferring Files with D-RATS

Under the **File** pull-down menu, there are three different options for initiating file transfers: **Broadcast Text File, Send File** and **Send Image.** When clicked, these three options allow you to browse to where the file to be transmitted is located, so the file to be sent can be selected.

Use **Broadcast Text File**, when you want to transmit a pre-formatted text message that you have created. As the name suggests, this is not a true file transfer because the file is being sent and received as if it was text data that was typed. You can create this file with any text editor, such as Windows Notepad or Word pad. You don't want your message to contain hidden formatting control sequences that are inserted by word processors. If you do use Microsoft Word, or some other word processor, save the file as a .txt file.

Send File allows you to actually transmit files that will be received as a file by the receiving station. When the file comes in, D-RATS will place it in the location specified by the **Paths**, **Download Directory** box found under the **Preferences** section of the **Config** window. If you send files on a regular basis, press your computer's **F1** key as a shortcut to using the **Send File** option on the **Files** menu.

The **Send Image** selection, true to its name, is used to send .jpg .gif and other types of image files. Be careful with sending images, as files can be quite large and may be unsuitable for transmission via D-STAR's low-speed DV mode.

File Transfer Problems

If text and quick messaging are working fine, but you are having difficulty transfering files, you might try the following to solve the problem.

Data Block Transfer Problems: Best throughput performance will be had with larger block sizes. If you are having trouble completing file transfers, a high block size may prevent you from getting blocks through. Reducing the block size reduces the odds of incurring errors within the block, which may reduce file transfer difficulties.

To change the default block size select **Main Settings** under the **File** pull-down menu, and then select **Radio, Transfers,** on the **Config** screen. The program default block size is 512, which is relatively large. To see if smaller block sizes can alleviate block transfer problems, you might try block sizes of 256 or 128.

Mixing File Transfers with voice: This is from the D-RATS web page. "Because of the effect timing has on the ability to detect failed or missing blocks and retry, interrupting a file transfer can cause problems. Talking on the digital voice side while doing a file transfer will prevent the proper negotiation from happening in a timely manner, and may cause the transfer to time out. Try to avoid talking while sending a file if possible."

Using and Creating D-RATS Forms

There are a number of pre-made forms available. To view available forms, click on the **File** pull-down menu and select **Manage Form Templates**, which displays all available forms. Click on one of the forms in the list and then click on **Edit**, which displays the **Form builder** window. The **Form builder** window allows you to create or edit forms. Click on **Preview** to view the selected form. After examining the forms, click on **Cancel** twice, and then click on **Close** to return to the main **D-RATS** window.

To select a pre-made form, so it can be filled out, from the main **D-RATS** window select the **Form Manager** tab and click on **New,** which displays the **Choose a form** window. Select one of the forms using the pull-down box and then click **OK** to display the form. Now it can be filled out and then: saved, sent, exported to a file folder, or printed.

If you wish to create your own forms, under the **File** pull-down menu, use the **Manage Form Templates** button to access the **Form Manager** window. From this screen you can either edit and rename an existing form, or build one from scratch. To learn how a form is created or edited, click on one of the existing forms to highlight it and click on **Edit** to open it. So you can freely experiment without modifying the original form, use the **Form builder** window and change the name in the **ID** box to something else and **Save** it as a test form. Now using this as a test form, closely examine its **Form Elements**, modifying them and adding new ones to learn how the form was created. Periodically use the **Preview** button to evaluate your modifications. After some experimentation you will learn how to edit and create your own forms.

Stations and Sessions Tabs

The **Stations** tab allows you to see the call signs, and GPS position, if sent, of stations that have been received. Stations can be deleted from this list via the **Remove** and **Clear All** buttons. Clicking on **Address** places the station's call sign in the **Send** text box. When you click on one of the stations in the list, it becomes highlighted, press **Forget** if you wish to reset the Last Seen time stamp to **Never**. If you press **Reset**, a selected station's **Last Seen** time stamp is set to the current time.

The **Sessions** tab can provide current status of transfers in progress and other information. When transferring files, it's instructive to see the file transfer's progress displayed on this page, especially if you are having file transfer difficulties.

Other D-RATS Capabilities

The above sections describe the main features and functionality of the D-RATS application. However, there are other features that advanced users may find of interest.

Ratflector: As described on the D-RATS web page, **http://d-rats.danplanet.com/wiki/Ratflector** the Ratflector is available for testing and asking questions of users that may be connected. Accessed via the Internet, the Ratflector is a non-RF connection that allows you to test and experiment with the various D-RATS modes of operation. Instead of routing data out a COM port to the radio, it is sent over the Internet. You can do just about everything that you would normally do over the air, but in this case the data packets are transferred via the Internet. Using the Ratflector, you could exchange data with a friend, or use two computers side-by-side to simulate two stations intercommunicating.

Internet email: D-RATS can be used to send and receive email to and from the Internet. A D-RATS station can be set up as an email server, so that stations without a direct Internet connection can still receive and send email messages. Emails are created and retrieved via the D-RATS **Form Manager**. Further information can be found on the D-RATS web page: **http://www.d-rats.com/wiki/InternetEmail**

TCP Forwarding: From the D-RATS web page: "D-RATS has generic TCP forwarding support. This means you can pipe some TCP traffic across the radio to a remote station. This may be useful for bridging SMTP traffic to a remote location or for doing POP3 for remote mailbox access. In it's native form, this will be rather slow, given that most of the protocols likely to be used in this fashion are very chatty and depend on a lot of back-and-forth handshaking to complete. While you should be prepared for this, the functionality should still be quite useful." Further information can be found at: **http://www.d-rats.com/wiki/TcpForwarding**

D-RATS Repeater Operation: D-RATS also has a repeater program, which can be used as a repeater with multiple radios, or as a network proxy (instead of or in addition to the repeater function). Further information can be found at: **http://www.d-rats.com/wiki/Repeater**

Connecting D-RATS to a DPRS Interface: Two separate pieces of software, *D-RATS* and *DPRS Interface* developed by Pete Loveall AE5PL, can be configured to operate simultaneously across a D-STAR radio's low speed DV channel. DPRS Interface is designed to pass GPS-A DPRS messages received from a D-STAR radio's low speed DV channel and pass them on to the APRS-IS network where it is combined with APRS and DPRS traffic world-wide. As many internet connected D-STAR repeaters already have an operational DPRS > APRS-IS interface, the D-RATS interface is only intended for use in cases where that connectivity is not available. Further information can be found at:
http://www.d-rats.com/wiki/DPRS-Interface

Chapter 9: **DV Dongle, D-STAR Adapter**

Besides being a solution for hams that don't have local access to a gateway equipped D-STAR repeater, the DV Dongle is fantastic for quickly and easily connecting to gateways and reflectors all over the world. It's especially suited for situations when you are out of town and want to get back in to your local repeater.

Developed by Robin Cutshaw, AA4RC and Moe Wheatley, AE4JY, the DV-Dongle provides an alternative method for participating in D-STAR communications without using a radio.

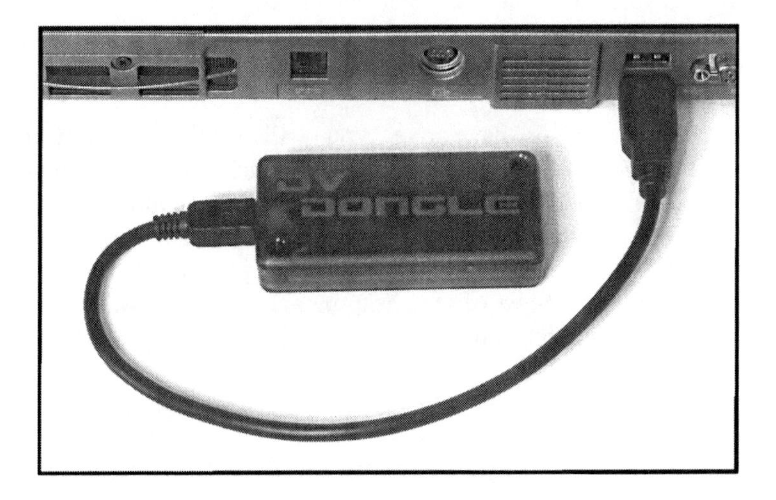

Housed in a small semi-transparent blue plastic enclosure, the Dongle connects to your computer's USB port via a short interface cable. Since world-wide D-STAR communications are facilitated by sending digitized voice packets over the Internet, the only thing a PC needs to be able to participate in D-STAR communications is some suitable software and a means of encoding and decoding D-STAR's digital voice streams. Taking advantage of a PC's high speed USB 2.0 data port, circuitry inside the Dongle uses an AMBE digital voice CODEC to encode and decode voice data streams to the D-STAR protocol. Robin Cutshaw's, *DV Tool* software than takes care of transporting these data streams back-and-forth over the Internet to remote gateways and reflectors.

Computer System Requirements

The DV Dongle application needs a relatively fast computer to run properly. It's recommended that you have a 2 GHz or faster machine, and turn off any compute intensive applications to prevent data overruns between your computer and the Dongle. The Dongle's red LED flashes whenever overruns occur. I first tried installing on an older 1 GHz system, and performance was unsatisfactory. Operation on my Intel Dual Core 1.6 GHz laptop was quite usable, with infrequent overruns.

In addition to a fast computer, you also need a high speed Internet connection. Dial-ups will not work. The Internet service that comes bundled with my cable TV subscription worked fine.

If you don't already have one, you will need a headset and microphone that is compatible with the audio jacks on your computer, typically 3.5mm (1/8") audio jacks. There are a lot of inferior quality headset / mic products being sold for the mass-consumer PC market; for optimum voice reception and transmission you will want to get a higher quality unit. Most of the cheap units have small ear cuffs that don't block out ambient noise and quickly become uncomfortable to wear.

Installing the DV Dongle Software on Your Computer

After purchasing a Dongle, the latest version of the *DV Tool* software is downloaded from the DV Dongle web page.
http://dvdongle.com/DV_Dongle/Home.html

Clicking on the **Installation** tab accesses installation instructions and sample screen shots. Instructions and different software downloads are provided for the Microsoft Windows, Mac OS X Leopard, and Linux operating systems.

Note that on Windows systems, inserting the Dongle will bring up the *Found New Hardware Wizard* to assist you with installing the required USB drivers. Installation of USB Drivers on Microsoft Windows is a two-step process, bringing up the *Found New Hardware Wizard* two times. Make sure you complete both installs.

After installing the USB driver, you can download the *DV Tool* software application to a suitable place on your computer. The web site instructions suggest you download and extract the files to your desktop. Preferring not to do that, I downloaded them to a *DVTool* sub-folder I created in my *Program Files* directory, which is where I keep all my other applications.

After extracting the files by unzipping them to whatever location you prefer, the application is started by double-clicking or opening the *DVTool.jar* file found in the *DVTools* folder. Rather than having to open the folder whenever you want to run the program, you might want to put a shortcut to the program on your desktop or in the PC's START menu.

On Windows, to place a short cut on the desktop, right click on the *DVTool.jar* file and select **Send to Desktop** from the pull-down window that pops up. To insert a shortcut into the **START**, **All Programs** menu, drag the *DVTool.jar* file to the START button; wait until the **All Programs** option shows, and then drop/place the shortcut where you want it to appear in the list of programs.

Selecting the DV Tool COM port and Audio Devices

On the DV Dongle home web page; select the **Using the DV Dongle** tab to access the operating instructions. Before starting the *DVTool* program, make sure the DV dongle is plugged in. Upon starting the program verify that the serial port showing in the **DV Device** window is correct. If not, select the correct one from the pull-down list.

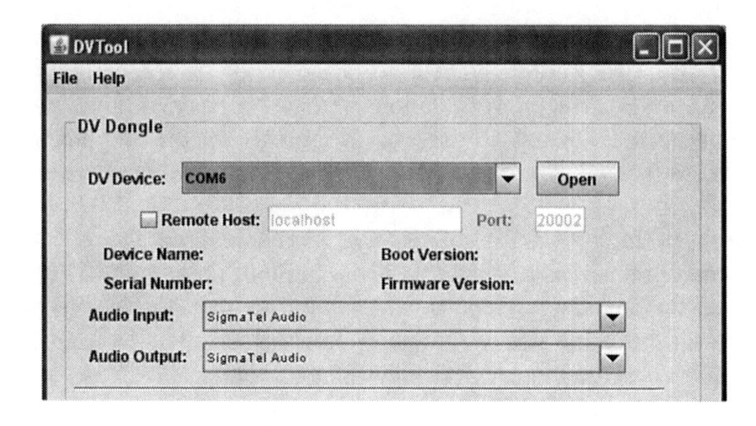

Next verify that the Audio Input and Output devices selected are the ones you want to use. If unsure, leave the default settings for now, you can come back to this and try other settings while doing **Audio Loopback** testing.

Click on **Open** to establish communication with the DV Dongle hardware. If all is well, you should see the **Device Name, Boot Version, Serial Number** and **Firmware Revision** fields filled in as shown below. If communication was not established, you may have the wrong COM port selected. Try a different setting.

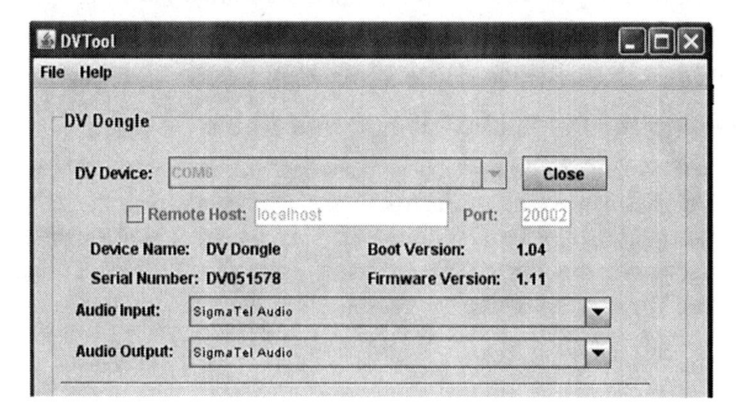

Setting the Headset and Microphone Audio Levels

Open your computer system's Volume Control settings window, usually accessed by right clicking on the speaker icon found in the system tray, at the bottom right of the computer screen.

Before setting the record level, you should first set the headset volume level for comfortable monitoring of other D-STAR station's transmissions. Once that is set, you can then set your microphone recording level to match received D-STAR signal audio levels. To do this we need to monitor some D-STAR activity.

Click on the **Connect to Gateway** button and select a D-STAR gateway with some traffic. This could be your own local gateway or a reflector, or a distant repeater. Next select the repeater's **Module,** if just looking for traffic, select the asterisk ∗ to be able to hear traffic on any of the modules. If you have not already done so, enter your own call sign in the **Callsign** box. By the way, even though you don't transmit, your call sign will be picked up by the D-STAR system and displayed as monitoring the selected repeater or reflector.

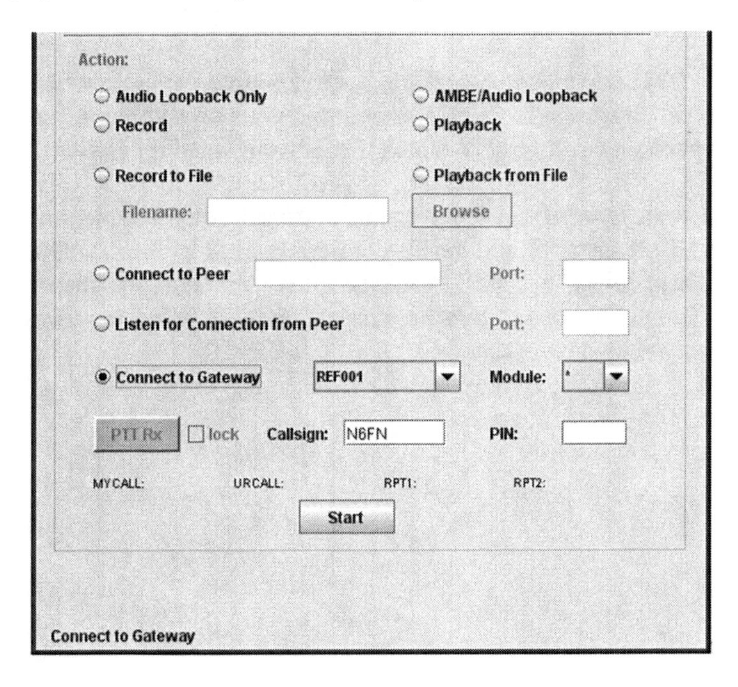

Once the call sign of the repeater or reflector to be monitored has been set, and your own call sign has been entered, click on the **Start** button to begin monitoring. If you don't hear any traffic on the selected repeater, you can access the last heard list at **http://dstarusers.org/lastheard.php** to find repeaters with some activity.

Monitor a few stations, adjusting the volume control for a comfortable listening level. As with conventional FM repeaters, you may notice volume level differences between received stations. If necessary, select a compromise volume setting that allows all stations to be comfortably heard.

Once your listening volume has been set, press **Stop** to disconnect from the D-STAR system so that the microphone recording level can be set. To access your computer's microphone level control, select **Properties** under the **Options** pull-down of the **Volume Control** window. This brings up the **Properties** window where you can select the **Recording** button, which in-turn displays the **Recording Control** window.

On the **DVTool** window, select the **Audio Loopback Only** function and then press **Start** again. While listening to your own voice, adjust the Microphone level control to match the listening level set earlier.

If you want to verify hear how your digitized voice will sound over the D-STAR network and verify the performance of the DV Dongle, press **Stop**, select the **AMBE/Audio Loopback** function and then press **Start** again. Here it will be a bit easier to monitor your own voice, as it is delayed by a few seconds.

DV Dongle Operation

In setting the volume levels, we have already covered how to monitor a D-STAR node. To transmit, just press and hold the **PTT / Rx** button. If it is grayed out preventing transmission, and the asterisk ∗ has been selected as the **Module**, you need to select a valid module, either A, B or C. Be aware, that at least on Version 1.11 of the *DV Tool* software that I was running, that it is possible to select a module that is not physically present at the repeater site. In that case, I don't know where your transmission actually goes.

The **PTT / Rx** button can also be grayed out preventing transmission if you are not registered for gateway transmission. Some gateway administrators may use the **PIN** field as a means of providing access to users not registered in the D-STAR gateway system. This field is normally left blank.

The lock function box can be selected if you wish to change the **PTT / Rx** button to operate as a toggle for transmit and receive. When set, clicking once enables transmission, clicking a second time returns to receive.

You may notice, that some repeater nodes do not appear in the **Connect to Gateway** selection box. This can be for a variety of reasons, as only gateways that are currently running the required DV Dongle access software and have their IP ports properly forwarded will be displayed.

Besides being able to select gateways via the pull-down menu, you may find it faster to type in a desired gateway's call sign. By clicking on the **Connect to Gateway** box and quickly typing you can enter a desired repeater's call sign. However, if you hesitate, only a portion of the call sign is captured, which selects the first call that matches the accepted portion.

Connecting to Repeaters Linked to a Reflector

If you connect to a repeater that is currently linked to a reflector, you will only be able to communicate with stations directly connected to the repeater. To be able to contact stations on the Reflector side of the link, you need to disconnect from the repeater and connect to the reflector that the repeater is linked to. Then you will be able to communicate with any station or repeater linked to that reflector.

Dongle LED Status Indicators

Four LED indicators, visible through the semi-transparent plastic case communicate the DV Dongle's operational status. A blue LED flashes whenever data is being transmitted from the computer and a flashing yellow LED indicates that data is being transmitted back to the computer. A slowly pulsing green LED indicates that the device is idle. More importantly for assessing performance, the red LED indicates either data overruns or under runs between the computer and the Dongle. Occasional overruns aren't too bad, but frequent flashing of the red LED indicates a slow computer. Shutting down other programs may help; otherwise you will need a faster system.

Installation Problems

A variety of problems can occur that will prevent DV Dongle operation. One of them is not having a sufficiently current version of *Java* running on your computer. If that is the case, the installation instructions found on the DV Dongle web page guides you through getting the latest *Java* updates downloaded.

Firewalls protecting your computer, network or router may prevent the program from communicating over the Internet. Other problems may arise with getting volume controls set, or being able to record from the microphone.

If you are having difficulty, the *Yahoo DV Dongle* group found at **http://groups.yahoo.com/group/DVDongle** may be able to provide assistance. With hundreds of postings covering just about any type of problem you can imagine, you should be able to find one that matches your problem. If you can't find a solution in the threads that are already posted, you might try registering and asking a question of your own.

Appendix: **D-STAR Web Pages**

Here are additional D-STAR web pages that you might find useful.

D-STAR Users, Source for D-STAR Information

http://dstarusers.org/lastheard.php has last heard lists, D-STAR repeater directory, gateway registration links and links to D-STAR accessory hardware and software and more.

D-STAR Information, FAQ's and More

http://www.dstarinfo.com/default.html D-STAR news, applications, links, nets, reflectors and more.

K5TIT The Texas Interconnect Team

http://www.k5tit.org/ Repeater group in Texas, lots of useful information.

Instructions for Making Icom Radio Interface Cables

http://epaares.org/dstar/icom_cables.htm GPS interface and PC programming cable-wiring diagrams for home made cables.

WY1U's D-STAR Repeater and Reflector Finder

http://home.comcast.net/~timmik/dstarsearch.html good for finding repeaters in different parts of the world.

APRS-IS / DPRS Information

http://www.aprs-is.net/ operating information, APRS-IS activity, APRS / DPRS conversion, software, specifications and more.

Enabling GPS Position Reporting through D-STAR

http://nj6n.com/dstar/dprs/ translating Icom GPS data to be compatible with APRS. Includes sample APRS radio setup guides.

Yahoo D-STAR Forum / Group

http://groups.yahoo.com/group/dstar_digital/ D-STAR discussion group, problem solving, and other topics

Icom Tech Support Knowledge Base

http://www.icomamerica.com/en/support/kb/Default.aspx
Select the following pull-downs for all D-STAR articles.
Product Line: Amateur
Product Class: D-STAR
Model: General
Balance of fields leave blank, Press **Search**

d*Chat Communications Software Home Page

http://nj6n.com/dstar/dstar_chat.html Feature list, software download, installation instructions.

D-RATS Communications Software Home Page

http://d-rats.com/ Feature list, screenshots, software download, installation instructions, user forum, FAQs.

D-StarCom Icom .icf Radio File Conversion Utility

http://www.d-starcom.com/ Program for converting Icom .icf files from one radio type to another.

jFindu Home Page

http://www.jfindu.net Last heard lists, D-STAR Repeater and DPRS activity, D-STAR repeater locator map and more.

DV Dongle Home Page

http://dvdongle.com/DV_Dongle/Home.html Technical info, Installation instructions, Software screenshots, FAQs

DV Dongle Yahoo Support Group

http://groups.yahoo.com/group/DVDongle DV Dongle discussion and support.

RT System's Programming Software

http://www.rtsystemsinc.com/index.cfm RT System's home page. Radio programming software and cables.

Made in the USA